"This innovative and engaging anthology on women and nature reveals the ongoing relevance of ecofeminism in today's global world by emphasizing postcolonialism, ecocriticism, queer ecology, animality, and feminist materialism. Anyone interested in the nuances and complexities of the women–nature connection across histories, belief-systems, and regions will want to buy this book."

— *Carolyn Merchant of the University of California at Berkeley, U.S., has written on the connections between ecofeminism and feminist theory and is the author of* Earthcare: Women and the Environment, *among other books.*

"The myriad ways that Earthly bodies—both human and non-human—continue to be bound by structures of patriarchy and domination requires sustained analysis. This transnational, transdisciplinary volume brings the lens of ecofeminism to bear on timely topics, including transgender studies, animal studies, and the new materialism."

— *Elizabeth Allison is the Program Chair of Ecology, Spirituality, and Religion at the California Institute of Integral Studies, U.S.*

"This fresh and exciting collection identifies privileges and invisibilities overlooked in earlier ecofeminist thinking. Authors call for ethical self-reflexivity and deep questioning of heteronormative assumptions reflecting a wide range of interdisciplinary, postcolonial, and cross-cultural perspectives. From ecosickness narratives to borderlands ecofeminism, this set of papers provides a rich and timely offering by deeply thoughtful scholars across the globe."

— *Stephanie Kaza, Professor Emerita, University of Vermont, U.S.*

Women and Nature?

Women and Nature? Beyond Dualism in Gender, Body, and Environment provides a historical context for understanding the contested relationships between women and nature, and it articulates strategies for moving beyond the dualistic theories and practices that often frame those relationships.

In 1974, Françoise d'Eaubonne coined the term "ecofeminism" to raise awareness about interconnections between women's oppression and nature's domination in an attempt to liberate women and nature from subordination. Since then, ecofeminism has attracted scholars and activists from various disciplines and positions to assess the relationship between the cultural human and the natural non-human through gender reconsiderations. The contributors to this volume present critical and constructive perspectives on ecofeminism throughout its history, from the beginnings of ecofeminism in the 1970s through to contemporary and emerging developments in the field, drawing on animal studies, postcolonialism, film studies, transgender studies, and political ecology.

This interdisciplinary and international collection of essays demonstrates the ongoing relevance of ecofeminism as a way of understanding and responding to the complex interactions between genders, bodies, and the natural environment. It will be of great interest to students and scholars of ecofeminism as well as those involved in environmental studies and gender studies more broadly.

Douglas A. Vakoch is President of METI International, a nonprofit research and educational organization devoted to Messaging Extraterrestrial Intelligence (METI) and supporting the sustainability of human culture on multigenerational timescales, which is essential for long-term METI research.

Sam Mickey is Adjunct Professor in the Theology and Religious Studies Department at the University of San Francisco, U.S.

Routledge Environmental Humanities

Series editors: Iain McCalman and Libby Robin

The *Routledge Environmental Humanities* series is an original and inspiring venture recognising that today's world agricultural and water crises, ocean pollution and resource depletion, global warming from greenhouse gases, urban sprawl, overpopulation, food insecurity and environmental justice are all *crises of culture*.

The reality of understanding and finding adaptive solutions to our present and future environmental challenges has shifted the epicenter of environmental studies away from an exclusively scientific and technological framework to one that depends on the human-focused disciplines and ideas of the humanities and allied social sciences.

We thus welcome book proposals from all humanities and social sciences disciplines for an inclusive and interdisciplinary series. We favour manuscripts aimed at an international readership and written in a lively and accessible style. The readership comprises scholars and students from the humanities and social sciences and thoughtful readers concerned about the human dimensions of environmental change.

Women and Nature?

Beyond Dualism in Gender, Body, and Environment

Edited by Douglas A. Vakoch and Sam Mickey

Routledge
Taylor & Francis Group
LONDON AND NEW YORK

earthscan
from Routledge

First published 2018 by Routledge
4 Park Square, Milton Park, Abingdon, Oxon OX14 4RN
605 Third Avenue, New York, NY 10017

Routledge is an imprint of the Taylor & Francis Group, an informa business

First issued in paperback 2018

British Library Cataloguing-in-Publication Data
A catalogue record for this book is available from the British Library

Library of Congress Cataloging-in-Publication Data
A catalog record for this book has been requested

ISBN: 978-1-138-05342-7 (hbk)
ISBN: 978-0-367-15240-6 (pbk)

Typeset in Bembo
by Keystroke, Neville Lodge, Tettenhall, Wolverhampton

Contents

Acknowledgments

As co-editors of this anthology, we must express thanks to the contributors, for without their thoughtful work this volume would not exist: Karen Ya-Chu Yang, Luca Valera, Anja Höing, Stephanie Baran, Anja Koletnik, Adriana Teodorescu, Denise Mitten, Chiara D'Amore, Valerie Padilla Carroll, Shlomit Tamari, Christina Holmes, Sreejith Varma, Swarnalatha Rangarajan, Reena Dube, and Izabel F. O. Brandão. We are grateful for their perspectives and their contributions to the ongoing development of ecofeminism. We would like to thank Rebecca Brennan, Kelly Watkins, and everyone else involved with the editorial team at Routledge. We appreciate their support for this work and their expertise in ushering it into publication.

Doug Vakoch is grateful to the members of METI International's Board of Directors for their friendship and creative collaboration. It is rare to find a community that sees a natural link between sustainability and the search for life in the universe, as reflected in a strategic plan that affirms the "ways that ecofeminism can provide insights into fostering environmental sustainability on multigenerational timescales." Doug especially thanks Harry and Joyce Letaw for their ongoing intellectual and financial contributions to METI. For creating new publishing opportunities for scholars working at the interface of literature and environment, he thanks the members of the Advisory Board of Lexington Books' Ecocritical Theory and Practice Series. Finally, to his wife, Julie Bayless, Doug is grateful in more ways than he can or will share here.

Sam Mickey would like to extend particular thanks to family, friends, students, and colleagues who support this work, including the University of San Francisco, where Sam teaches in the Theology and Religious Studies Department and on Environmental Studies program. The learning community at USF is welcoming and empowering for research committed to more peaceful, just, and sustainable relationships among genders, bodies, and environments. Sam would also like to thank the ecofeminists with whom he has been in conversation personally and professionally, including particularly informative and compelling dialogue with Kimberly Carfore, Lois Lorentzen, Catherine Keller, Whitney Bauman, and Donna Haraway.

Finally, acknowledgments must also be given to the nonhuman bodies and environments that have been part of this book—the life, land, air, and water that has nourished and inspired the contributors and provided materials for the publication and distribution of this book.

Douglas A. Vakoch and Sam Mickey

Contributors

Stephanie Baran is a Ph.D. student and lecturer at the University of Wisconsin-Milwaukee, U.S. Previously, she was an adjunct lecturer at Kankakee Community College and Prairie State College, teaching several semesters of introductory sociological theory and race and ethnicity courses. She received her MA in Sociology from DePaul University in Chicago, U.S. She is a student, researcher and activist and is an avid student of Marxist theory. Her academic interests include studying theory, racism, sociology of the body, feminist and Marxist theory. Her master's research paper, "Parsing White Supremacy: An Exploratory Study of Political Thought and Beliefs," was published with Heathwood Institute and Press in December 2014. She also published a personal essay entitled "Scars, Womanness, and Motherhood" to Feministing.com in May 2014. And most recently, "Access to Parlay: Freedom of Speech" was published to Heathwood Institute and Press in February, 2015. Her hobbies include being a fiery feminist, a gay and trans ally, and working toward total social justice. She currently resides in Milwaukee with her cat, Harley.

Izabel F. O. Brandão is Professor of Literatures in English, and Contemporary Brazilian Women Writers at the Federal University of Alagoas, Brazil. Her publications include *A imaginação do feminino segundo D. H. Lawrence* ("The imagination of the feminine according to D. H. Lawrence," Edufal, 1999), and books on Brazilian women writers and feminist literary criticism from an interdisciplinary perspective on the body, such as *O corpo em revista* ("The body under scrutiny," 2005), *Gênero & outros lugares* ("Gender & Other Places," 2009, in collaboration). She is currently editing (in collaboration) a feminist anthology in translation (*Traduções da cultura:* perspectivas crítcas feministas *1970–2010* ["Translations of Culture: Feminist Perspectives 1970–2010"], forthcoming), and a book provisionally entitled *Encontros feministas com D. H. Lawrence* ("Feminist encounters with D. H. Lawrence," forthcoming). She is also a poet and has three books published *Espiral de fogo* ("Fire Spiral," 1998), *Ilha de olhos e espelhos* ("Island of eyes and mirrors," 2003) and *As horas da minha alegria* ("The hours of my joy," 2013).

Chiara D'Amore holds a Ph.D. in Sustainability Education from Prescott College, U.S., is the founder and director of Columbia Families in Nature, and is a Toyota-Audubon TogetherGreen Fellow. Her research is focused on understanding the effects of time spent in nature on personal, social, and ecological well-being. She recently received the Howard County Celebrating Excellence, Successes in Bloom award for her work engaging mixed ages in educational nature experiences. She currently serves on the Board of Directors for World House Community Supported Education and the Audubon Society of Central Maryland. She holds a Master of Science in Environmental Science and Engineering and works as an environmental consultant, focusing on voluntary environmental program design, implementation, and evaluation. Recent publications include: "Thriving Through Nature: Fostering Children's Executive Function Skills," a leadership series paper with the Children & Nature Network (2015); "Nurtured Nature: The connection between care for children and care for the environment" in *Pedagogies of Kindness and Respect: On the Lives and Education of Children* (2015); "Cultivating Connection and Care – The Case for Family Nature Clubs" in the *Journal of Sustainability Education* (March 2015); and "Family Adventures in Nature Club Guidebook" for the Izaak Walton League of America (2015).

Reena Dube is an Associate Professor of Global Literature and Film and has published two books: *Satyajit Ray's "The Chess Players" and Postcolonial Theory: Culture, Labour and the Value of Alterity* (2005) and a coauthored book, *Female Infanticide in India: A Women's Cultural History* (2005). She has published articles on film, lyric poetry of Meera, Bhats and Charans, subaltern resistance, in journals like *Postscript, Boundary 2, Works and Days*, and *Subaltern Studies*. She has been the editor of the journal *Studies in the Humanities* published by the English Department at the Indiana University of Pennsylvania since 2010. Currently she is in the process of putting together an edited book on cityscapes.

Anja Höing studied English and Biology at the University of Osnabrück, Germany. She obtained her Master of Education (M.Ed.) in 2011, and is now employed as a research assistant at the institute of English and American Studies at the University of Osnabrück, where she is working on her Ph.D. on religion and culture in English animal stories, and teaches a range of undergraduate courses on British literature. Her main research interests lie in the fields of animals in children's literature, representations of nature and ecosystems and environmentalism in literature. She presented her research at several conferences, such as "Cosmopolitan Animals" (2012), "Reading Animals" (2014), the ASLE "Green Knowledge" conference (2015), and the 1st and 2nd "Global Conference: The Animal and Human Bond" by Inter-Disciplinary (2014 and 2015). One of her conference papers on the topics of Richard Adams's *The Plague Dogs* has since been published in the edited volume *Who's Talking Now? – Multispecies Relations from Human and Animals'*

Point of View and other papers are soon to follow. She is also a member of the ASLE-associated postgraduate workshop "Environment, Literature, Culture" (ELC).

Christina Holmes is an Assistant Professor of Women's, Gender, and Sexuality Studies at DePauw University in Indiana, U.S., where she teaches interdisciplinary courses on ecofeminism and environmental justice, women of color in the U.S., and transnational feminism. She completed her Master's degree at Oxford University and earned her doctorate in Women's, Gender, and Sexuality Studies at the Ohio State University. Her current research explores the intersections between ecofeminism, new material feminisms, and Chicana studies—a manuscript on this topic won the National Women's Studies Association First Book Prize and is forthcoming with the University of Illinois Press (2016). While working on a draft of this chapter, Christina was a Visiting Scholar at the Department of Gender and Cultural Studies at the University of Sydney.

Anja Koletnik is a transfeminist queer activist and freelance academic from Slovenia. Anja holds a B.A. in Social Work from University of Ljubljana, Slovenia, and an M.A. in Gender Studies from Central European University, Hungary. They are particularly interested in the fields of transgender studies, queer theory, fat studies, embodiments and feminist new materialisms. They have presented papers at "(Im)possibly Queer International Feminisms," the 2nd International Feminist Journal of Politics Conference, sponsored by Centre for Advanced international theory, University of Sussex, "Body Discourses/Body Politics and Agency" held by The Gender Research Office, University of Vienna and "New Materialist Politics and Economies of Knowledge" held by the Faculty of Law, University of Maribor. Anja has various years of experience in LGBTQ+ and feminist activism and is the founder and director of Slovenia's first transgender specific non-governmental organization Institute Transfeminist Iniciative TransAkcija. Their current work is mostly focusing on legitimization and recognition of the genders, gender identities and transgender/gender variant identities within LGBT activism, feminisms and academia in Slovenia.

Sam Mickey is Adjunct Professor in the Theology and Religious Studies Department and the Environmental Studies program at the University of San Francisco in San Francisco, California, U.S. He is the author of multiple books, including *Coexistentialism and the Unbearable Intimacy of Ecological Emergency* (2016), *Whole Earth Thinking and Planetary Coexistence: Ecological Wisdom at the Intersection of Religion, Ecology, and Philosophy* (Routledge, 2015), and *On the Verge of a Planetary Civilization: A Philosophy of Integral Ecology* (2014). He is also co-editor (with Sean Kelly and Adam Robbert) of *The Variety of Integral Ecologies: Nature, Culture, and Knowledge in the Planetary Era* (SUNY, 2017). He blogs regularly at BecomingIntegral.com.

Denise Mitten is the chair of the Sustainability Education Program at Prescott College, U.S. She is internationally recognized for her innovative scholarship in outdoor and environmental pedagogy, ethics, and gender, serves as a reviewer and/or board member for academic journals such as the *Journal of Experiential Education* and *Journal of Sustainable Education*. She has published more than 30 papers and co-authored N*atural Environments* and *Human Health* and co-edited *Experiential Education Theory and Practice*. In 2015 Denise received the Distinguished Researcher Award for her creative and innovative impacts to research within the experiential education community. A founder of "Woodswomen" in the U.S., she is a world-renowned adventure guide (Mt. McKinley/Denali, Himalayas) and an early promoter of ecotourism. Through her leadership, Woodswomen, Inc. offered a venue for women to heal in nature and develop outdoor living skills. Starting with the Girl Scouts in the 1960s, she has worked with many populations, including youth, nuns in recovery, survivors of abuse, and women felons, and has led trips from SCUBA to mountaineering, rafting, kayaking, rock climbing, skiing, cycling, and more. She combines the ethic of care with expert outdoor skills guiding people in ecopsychology and therapeutic experiences in nature.

Valerie Padilla Carroll is Assistant Professor of Gender, Women, and Sexuality Studies at Kansas State University, U.S., where she teaches classes on feminist justice movements and environmentalism as well as classes that explore the intersections of gender, race, class, and nature in film and television. Her publications include, "Sustainable Universities: Rhetoric versus Facts," in *Environmental Policy is Social Policy – Social Policy is Environmental Policy: Towards Sustainability Policy* (2013) and the upcoming "Bridging the Radical/Mainstream Environmentalist Divide: Affect and Objectivity in the Writings and Interpretations of Aldo Leopold," in *The Aldo Leopold Reader* (2016). Her current research centers on histories of feminist and other radical back-to-the-land movements. She is preparing a book provisionally titled *Who Get to go Back-to-the-Land? The Gendered and Racialized Self in in Self-Sufficiency,* which will analyze 20th- and 21st-century popular culture manifestations of U.S. back-to-the-land movements. She is involved in developing Transition Towns and promoting permaculture in Kansas.

Swarnalatha Rangarajan is Professor of English at the Department of Humanities and Social Sciences, Indian Institute of Technology, Madras, India. She was a Fulbright Pre-doctoral fellow at Harvard University in 1999–2000 and was awarded her Ph.D. from the University of Madras in 2002. She is the founding editor of the *Indian Journal of Ecocriticism* (IJE) and has served as Guest Editor for two special issues on Indian ecosophy for *The Trumpeter*. She has published articles on ecocriticism in several journals and serves on the editorial board of *ISLE* (*Interdisciplinary Studies in Literature and Environment*), the prime journal in the domain of ecocriticism. She was awarded the Charles Wallace Fellowship at CRASSH (Cambridge

University) in 2013 to work on (co-edited) book projects titled *Ecoambiguity, Community, and Development: Toward a Politicized Ecocriticism* (2014) and *Ecocriticism of the Global South* (2015) published by Lexington Books. Swarnalatha Rangarajan also dabbles in creative writing and her short fiction has appeared in anthologies of publishing houses like Penguin, Zubaan, and Westland, to name a few. Her first novel, *Final Instructions*, which has a prominent ecosophical theme, was published in 2016.

Shlomit Tamari is a lecturer in Philosophy of Education at the department of Interdisciplinary Studies, Sapir Academic College, Israel. She was awarded her Ph.D. from Ben-Gurion University in the Negev in 2005. Her publications include two books and several articles: *Maurice Merleau-Ponty's Phenomenology of Perception: A Basis for Sharing the Earth* (2004, with Haim Gordon); *Maurice Merleau-Ponty and Environmental Education* (2008, Hebrew); "Active Environmental Citizenship at the Hebron-Bessor Drainage Basin" (2015, with Renana Ilan); "The Marcus Family Campus: New Design and Contemporary Architecture" (2014, with Inbal B. Gitler); "How Merleau-Ponty Can Provide a Philosophical Foundation for Vandana Shiva's Views on Biodiversity" (2011). In addition, Shlomit Tamari is an educator and environmental activist who led the Hebron Stream High Court petition and public campaign. She is an initiator and leader of the Forum for the Hebron-Bessor Drainage Basin and a member of the executive board of Sustainable Development for the Negev Organization.

Adriana Teodorescu has a PhD in Comparative Literature (2008-2011) and is a PhD Student in Sociology, Babeş-Bolyai University, Cluj-Napoca, Romania (2016-present). She is a member of the *American Comparative Literature Association* (ACLA), *The Association for the Study of Death and Society*, UK, and the *Athens Institute for Education and Research* (ATINER)—the Literature Research Unit and the Sociology Research Unit, Greece, and she worked as a Visiting Scholar at *The American University of Paris* in 2010. Her latest publications include "The Contemporary Imaginary of Work: Symbolic Immortality within the Postmodern Corporate Discourse," in *Postmortal Society. Towards a Sociology of Immortality* (2017). She has been co-organizer of the annual "International Conference Dying and Death in 18th-21st Century Europe" since 2010. In March 2016, she organized a seminar at Harvard University (*Interdisciplinary Perspectives on Death Representations in Literature: Epistemological, Social, Anthropological and Aesthetic Aspects*), as part of the ACLA's Annual Meeting. She co-organized "Death, Dying and the Disposal of the Body Conference: Eastern and Western Ways of Dying and Death" (DDD12). She is editor of *Death Representations in Literature: Forms and Theories* (2015), and co-editor of *Dying and Death in 18th-21st Century Europe* (2 volumes, 2011 and 2014). Her fields of interest are death studies, gender studies, sociology of ageing and old age, comparative literature and cultural studies.

Douglas A. Vakoch is President of METI International (http://meti.org), a non-profit research and educational organization devoted to Messaging Extraterrestrial Intelligence (METI) and offering programs to the public and to the scholarly community that foster increased awareness of the challenges facing our civilization's longevity, while encouraging individual and community activities that support the sustainability of human culture on multigenerational timescales, which is essential for long-term METI research. His edited books include *Ecofeminism and Rhetoric: Critical Perspectives on Sex, Technology, and Discourse* (2011), *Feminist Ecocriticism: Environment, Women, and Literature* (2012), and (with Fernando Castrillón) *Ecopsychology, Phenomenology, and the Environment: The Experience of Nature* (2014). He serves as general editor of the Ecocritical Theory and Practice series (https://rowman.com/Action/SERIES/LEX/ETAP), published by Lexington Books, an imprint of Rowman & Littlefield.

Luca Valera is Professor of Bioethics and Moral Philosophy at Pontificia Universidad Católica de Santiago de Chile. Luca completed his studies in Philosophy at Università Cattolica del Sacro Cuore in Milan, before undertaking a Ph.D. in Bioethics at Università Campus Bio-Medico di Roma. He has been Lecturer in Fundamentals of Anthropology and Ethics in the Faculties of Medicine and Engineering at the University Campus Bio-Medico in Rome, Lecturer in Human Ecology and Sustainability in the Faculty of Engineering at the University Campus Bio-Medico in Rome and Lecturer in Medical Anthropology in the Faculty of Medicine and Surgery at Università Cattolica del Sacro Cuore in Rome. His main interests are: Moral philosophy, Environmental Ethics, Philosophy of Ecology and Sustainability and Gender Philosophy. His latest publications are: *Arne Næss. Introduzione all'ecologia* (2015); *Educare al Ben-Essere. Lineamenti di etica* (Aracne, Roma, 2015, co-author M. T. Russo); *Ecologia Umana. Le sfide etiche del rapporto uomo/ambiente*, (2013); "Peter Singer and Food; Human Ecology and Food," in P. B. Thompson, D. M. Kaplan (eds.), *Encyclopedia of Food and Agricultural Ethics* (2014); "Post-Humanism: Still or Beyond Humanism," *Cuadernos de bioetica*, 25/3 (2014), 481–491; "Desarrollo humano sostenible: una visión aristotélica," *Isegoria. Revista de la filosofia moral y politica*, 51 (2014), 671–690 (co-author A. Marcos); "Un nuovo cancro per il pianeta? Natura, ambiente ed essere umano nell'etica ambientale contemporanea," *Teoria. Rivista di filosofia*, 2 (2014), 175–192.

R. Sreejith Varma is a Ph.D. student at the Department of Humanities and Social Sciences, Indian Institute of Technology, Madras, India, where he works on green subaltern writing in Malayalam. He holds M.Phil in Comparative Literature and M.A. in English from the University of Hyderabad. He has attended/presented papers in several national and international conferences on comparative literature and ecocriticism. His publications include "Comparative Literature and Translation: The Changing Contours of Relationship" in the journal *Scholar Critic* (2014), "Writing

Back: Narayan's *Kocharethi* as the First Adivasi Novel in Malayalam" in the journal *Luminaire* (2015) and a book chapter titled "The Possibilities of a River and a Dance: An Ecoethnographic Analysis of Kuttan Aarangottu Vayali's *Bhagavathy aattu*" (co-authored with Swarnalatha Rangarajan) for *Ecodocumentaries: Critical Essays*, edited by Rayson K. Alex and Susan Deborah (2016). His Malayalam-English translations of two short stories by Narayan, the first tribal fiction-writer from Kerala, have appeared in the journals *eDhvani* (2015) and *Muse India* (2015).

Karen Ya-Chu Yang is an Assistant Professor of the Department of English at Tamkang University, Taiwan. She received her Ph.D. in Comparative Literature from Indiana University–Bloomington in 2013. Her research interests include East–West comparative studies, postmodernism and post-colonialism, contemporary fiction and film, as well as feminist and ecological inquiries. She has received research grants from Chiang Ching-Kuo Foundation (2012–2013) and Taiwan's National Science Council (2014–2015). Some of her publications include "Prince of the Himalayas: A Reconstruction of Tibetan Cultural Identities through the Meeting of Tibet, Shakespeare, and China" (2010), "Pig-Women on the Meat-Market: Problems and Potentialities of Ecofemnist Hybridity" (2013), "Rewriting Canonical Love Stories From the Peripheries" (2013), "Passionately Documenting: Taiwan's Latest Cinematic Revival" (2015), and "Angels and Feathers: Transcorporeal Morphing in Angela Carter's Nights at the Circus" (2016).

Editor's foreword

Sam Mickey

This book grew out of a shared interest among the editors and contributors in the ongoing development of ecological feminism, or, as it is more commonly and succinctly referred to, ecofeminism. The chapters in this book present critical and constructive perspectives on ecofeminism throughout its history, from the beginnings of ecofeminism in the 1970s through contemporary and emerging developments in relationship to critiques it has received and other fields of study with which it intersects.

First finding explicit expression in the 1970s, ecofeminism began with various efforts to develop theories and methods for understanding and responding to the complex connections between gender roles, sexually differentiated bodies, and the life, land, air, and water that make up the natural environment. Focusing on intersections of humans and non-humans, ecofeminism thus stands in contrast to feminism, which views human issues in relative isolation from ecological concerns, and in contrast to ecological descriptions of non-human organisms and ecosystems. Ecofeminism accounts for the intimate intertwining of human and non-human ways of being. Accordingly, ecofeminists are often critical of the dualistic tendency to treat humans and non-humans as two mutually exclusive categories. Historically and presently, dualistic categories are used as justification for the aggrandizing of humans as the sole subjects in the universe and the subordination of non-humans as mere objects with no agency or intrinsic value. Not only is the human/nature dualism harmful toward non-human modes of existence, but it also has disastrous consequences for humans, as quickly becomes obvious to anyone who attempts to live without regular environmental inputs of food, clean freshwater, and breathable air.

Although the word "ecofeminism" is a singular noun, ecofeminism is not simple or homogeneous. It involves a dynamic diversity of perspectives, changing and developing over time while traversing academic fields and activist forces across multiple subject positions of race, class, gender, age, sexual orientation, religion, and nationality. This anthology reflects that diversity, both in the topics addressed in the chapters and in the backgrounds of the contributors, who represent a variety of nations, including Brazil, USA, Taiwan, India, Germany, Italy, Slovenia, Romania, and Israel. This anthology also reflects the shared commitment of ecofeminists to overcome the dualistic hierarchy that categorically separates humans from the natural environment. Moreover, the human/nature dualism intersects in various ways with many other dualisms.

One of the crucial insights of ecofeminists is that the human/nature dualism intersects with the dualism between men and women. Just as the human/nature dualism is used historically and presently as justification for human subjects exploiting the natural world as mere objects, the man/woman dualism is used as justification for exploitative attitudes and actions of men toward women. Subjectivity is attributed predominantly to maleness and masculinity, while femaleness and femininity are objectified. Racist and classist hierarchies follow a related logic whereby differences are reduced to dualistic hierarchies that grant subjectivity and agency to one pole of the dualism while the other pole is reduced to an object of control. Dualisms of whiteness/color and rich/poor are thus also a profound concern for ecofeminists, as are dualisms of reason/emotion, mind/body, human/animal, sameness/difference, cisgender/transgender, and self/other. Ecofeminists have been critiquing those dualisms and offering alternatives since the 1970s. However, much more work remains to be done. Those dualisms pervade modern societies, and their roots go deep into the history of civilization. The social system oriented around the rule of men (i.e., patriarchy) is thousands of years old. Dualisms are so deeply entrenched that it is going to take more than a few decades of ecofeminism to completely fix the problem.

The anthology aims to further the development of ecofeminist critiques of and alternatives to dualisms. This entails addressing the shortcomings of earlier articulations of ecofeminism, some of which tend to treat different dualisms as symmetrical to one another, as if the objectification of women by men is exactly parallel to the control of emotions by reason or the exploitation of the environment by humans. While dualistic hierarchies intersect, it is important to attend to the differences between them and not overemphasize their symmetry or continuity. This point has been noted by critics of ecofeminism and by ecofeminists themselves who embrace intersectional modes of analysis. It has also been noted in other fields of inquiry that address intersections of human and non-human modes of existence, including fields like political ecology, environmental ethics, animal studies, ecological literary criticism (ecocriticism), and environmental humanities. The chapters in this anthology provide updates for ecofeminism, responding to critiques of earlier versions of ecofeminism and articulating alliances between ecofeminism and other related fields of inquiry.

One could say that this volume is facilitating a return of ecofeminism. However, this is not precisely a come-back or resurgence. Ecofeminism never went away. Since its inception, it has been multiplying and diversifying in response to critical and constructive feedback and in response to the emergence of other disciplines oriented around human/nature entanglements. For instance, postcolonial studies and queer theory challenged the essentialism of much ecofeminism, but ecofeminism did not disappear. Instead, ecofeminists began rethinking their ideas of unchanging definitions or "essences" of women and nature. Understandings of essences that apply in all times and places were replaced with more nuanced understandings of the ways that definitions of women and nature are constructed and performed across different cultures, histories, and places. As ecofeminist understandings of women and nature have changed, new designations have emerged for studies of complex connections between gender, body, and environment. Ecofeminism is thus closely connected with areas of study such as feminist ecocriticism, queer ecology, and new feminist materialism.

The book is divided into four sections, beginning with an overview that includes an introductory chapter and an analysis of the beginnings of ecofeminism in the work of the French feminist who first coined the term, Françoise d'Eaubonne. The second section highlights the contested role of animality in women–nature connections. The next section explores different conditions under which women–nature connections are constructed, elucidating the variously harmful and beneficial consequences of such construction. The final section considers ecofeminist practices that facilitate mediation and reciprocity across differences. The book concludes with an Afterword that draws together contrasts and common themes from all the chapters.

Overall, this volume aims to further ecofeminist thought and practice while also re-establishing grounds for ecofeminism in light of its critical reception. The contributors facilitate dialogue between different areas of environmental humanities and gender studies, representing the diversity of contemporary voices in ecofeminism, and empowering discourses and practices that critique and present alternatives to the dualistic hierarchies that typically frame ecological and social relationships (e.g., masculine/feminine, reason/emotion, native/foreign, cultural/natural, self/other). This anthology demonstrates the ongoing relevance of ecofeminism for understanding and responding to the complex interactions of genders, bodies, and environments in the diverse contexts of a globalized civilization.

Part I
Overview

Introduction

Karen Ya-Chu Yang

In 1974, Françoise d'Eaubonne boldly coined the term "ecofeminism" in her foundational text *Le féminisme ou la mort*. She introduced the idea to raise awareness about interconnections between women's oppression and nature's domination in an attempt to liberate women and nature from unjust subordinations. Since then, ecofeminism has attracted scholars and activists from various disciplines and positions to assess the relationship between the cultural human and the natural non-human through gender reconsiderations. The coalition of women studies with ecocriticism has incited much controversy among feminists. After the women's suffrage movement's fight for political equality during the 19th and early 20th centuries, feminist projects of the 1960s expanded gender rights to social relations such as the workplace as well as to reproductive rights and problems of domestic abuse. Into the 1980s, an emerging new generation started to profess dissatisfaction toward the exclusive essentialism and oppressive universalism of previous feminist movements. Research from various fields of study, ranging from postcolonial and diasporic decentering attempts to poststructuralist and queer deconstructions, joined forces with feminist programs to destabilize and diversify the definition and fixity of binary oppositions and restrictive categorizations. As a result, the rise of ecofeminism or ecological feminism during the 1980s evoked much concern for fear of a fallback into dualistic oppressions by realigning women with nature, body, and the environment. For the movement's supporters, however, the realignment between women and nature is not an essentialist return to feminizations of nature or naturalizations of women that previous feminist movements have fought hard to dismantle; on the contrary, ecofeminism argues for critical deconstructions of historical, cultural, and social dominions of women and nature in favor of a more complex, diversified, and hybrid bridging of the relationship between women and nature.

Ecofeminism reconsiders nature and environmental issues from a feminist perspective to uncover the perseverance of androcentric dualist thinking in society. Earlier writings such as Susan Griffin's *Women and Nature: The Roaring Inside Her* (1978) traced the biased treatment of women and non-humans in Western history to depict women's alliance with nature as a result of the dominance of patriarchy and androcentrism. Into the 1980s and 1990s, scholars continued to attack oppressions of women and nature and called for their joint emancipation from Man's oppression and subordination (Shiva 1988; Sturgeon 1997; Sandilands 1999). To do so, ecofeminist inquiries took to criticizing and demolishing the sovereignty of Western dualistic thinking. Some representative works include Elizabeth Fishers's *Woman's Creation: Sexual Evolution and the Shaping of Society* (1979), Marilyn French's *Beyond Power: On Women, Men, and Morals* (1985), Carol J. Adams's *The Sexual Politics of Meat* (1990) and her co-edited books with Josephine Donovan such as *Animals and Women: Feminist Theoretical Explorations* (1995), Greta Gaard's edited volume *Ecofeminism: Women, Animals, Nature* (1993), Val Plumwood's *Feminism and the Mastery of Nature* (1993), and Lynda Birke's *Feminism, Animals, Science: The Naming of the Shrew* (1994), Noël Sturgeon's *Ecofeminist Natures: Race, Gender, Feminist Theory and Political Action* (1997), and Catriona Sandilands's *The Good-Natured Feminist: Ecofeminism and the Quest for Democracy* (1999).

Through reflective re-associations between women and nature, ecofeminism aims to destabilize hierarchal dualisms where men rule as reason, subject, and master colonizing women as nature, object, and slave. This calls for an "anti-dualist ecological feminism" (Plumwood 1993, 40) so women can "move to a further stage in their relations with nature, beyond that of powerless inclusion in nature, beyond that of reaction against their old exclusion from culture, and towards an active, deliberate and reflective positioning of themselves *with* nature against a destructive and dualising form of culture" (39). Bringing women back in contact with nature does not suggest reinstalling fixed binaries; rather, such attempts aim to directly challenge the oppressiveness of dualistic hierarchies dominating the core of Western philosophy. Critical attention to the intersection between women's oppression and environmental crises raises awareness to operating power politics which involve the sufferings of various groups marginalized as subordinate Others. Like Karen Warren contends, "Ecofeminist philosophy extends familiar feminist critiques of social isms of domination (e.g., sexism, racism, classicism, heterosexism, ageism, anti-Semitism) to nature (i.e., naturism)" (1997, 4). The way we treat nature is directly related to and reflected in the way we treat each other (O'Brien and Cahn 1996; Gaard 2001; Beyer 2014). The goal is to re-access human's relationship with our material bodies and environment in a non-dominating manner so as to evoke more dynamic interactions with and between other human and biological communities.

Into the new millennium, with the emergence of environmental humanities, inquiries into women–nature relationships continue to broaden with diversity via approaches concerning ecocriticism, critical animal studies, queer theory, vegetarianism, biodiversity, new materialism, care theory, participatory epistemology as well as explorations from interdisciplinary, postcolonial, and cross-cultural perspectives. In underscoring nature and the body, the materialist turn of ecofeminist studies seeks to reconceive human and non-human relationships by looking into the dynamic interactions between material elements and social constructions. This material–discursive mangling aims at a "dynamic, nonessentialist and relational brand of materialist vitalism" (Braidotti 2014, 241).

As Serpil Oppermann remarks, "Material feminist epistemologies especially encourage studies of sociocultural, literary, and ethical dimensions of the new material paradigm, offering a compelling model which casts matter (all physical substances) and bodies (human and non-human) not as mere objects of knowledge, but as agents with vitality of their own, and as interrelated forces beyond human control, linking human corporeality with non-human life processes" (2013, 27). Material feminism conceives of our bodies as intimately connected to the environment, as seen with Stacy Alaimo's concept of "trans-corporeality," which envisions the movement across bodies as revealing "the interchanges and interconnections between various bodily natures" (2010, 212). Also devoted to deconstructing binaries, feminist care theories criticize universal judgments abstracted from detached reasoning, and instead practice participatory epistemology set in context to draw out intersections between care and justice, emotion and rationality, material and social/cultural/historical, as well as human and non-human. Acknowledging these "agencies of entangle-ment" generates the need for humans to develop feelings of care and sympathy toward the various existences on our planet (Barad 2007, 33). The past few years have witnessed an increase of collections which manifest interdisciplinary and international observations produced from various backgrounds. Some edited volumes of interest include *Women Writing Nature: A Feminist View* (Cook 2008), *Material Feminism* (Alaimo and Heckman 2008), *Species Matters: Humane Advocacy and Cultural Theory* (DeKoven and Lundblad 2012), *East Asian Ecocriticisms: A Critical Reader* (Estok and Kim 2013), *International Perspectives in Feminist Ecocriticism* (Gaard, Estok, and Oppermann 2013), *Ecofeminism: Feminist Intersections with Other Animals and the Earth* (Adams and Gruen 2014), and *The Rise of Critical Animal Studies: From the Margins to the Centre* (Taylor and Twine 2014), *A Political Ecology of Women, Water, and Global Environmental Change* (Buechler and Hanson 2015), *Contemporary Perspectives on Ecocriticism* (Phillips and Rumens 2015).

Adding to a more multifaceted exploration of women–nature relationships, this volume includes studies from various disciplines and perspectives to further contemporary ecofeminism's development beyond dualisms. The chapters apply different methods to dismantle hierarchal binaries of disseverance for the purpose of re-conceptualizing humans as dynamic beings of and participants in nature–culture environmental systems. As Stacy Alaimo, "I believe that neither a feminist retreat into nature where we pose as 'angel[s] in the ecosystem,' nor a feminist flight from nature is the answer. Instead, we must transform the gendered concepts – nature, culture, body, mind, object, subject, resource, agent, and others – that have been cultivated to denigrate and silence certain groups of humans as well as nonhuman life" (10). Interconnectedness and diversity constitute the essence of this interdisciplinary and international collection of essays. This volume joins together researchers from literary, media, postcolonial, sociology, environmental, gender and sexuality, psychology and health, as well as theoretical professions to offer varied approaches to the subject of women and nature. To forward the discipline of ecofeminist criticism, selected chapters offer re-examinations of prominent ecofeminist theories by François d'Eaubonne, Carol Adams, and Greta Gaard from fresh perspectives such as maternity arguments, transgender studies, and new materialism. In addition to research from the United States, this book also incorporates multiple case studies discussing women–nature challenges and activism in Israel, Mexico, and India. As a manifestation of scholars from various disciplines and national backgrounds, this collection serves as a collaborative contribution to generating more hybrid conversation regarding the global move beyond dualism in woman and nature interrelationships.

The collection begins with Luca Valera's chapter, which provides background discussion on core ideas of Simone de Beauvoir and François d'Eaubonne. In this opening chapter, Valera traces the roots of men's domination of women and nature to concepts of maternal dependency and care as well as to the major theorists of the Scientific Revolution who defined the modern age. Valera's chapter discusses the historical and practical justification of ecofeminism and the need to overcome dualistic thinking in order for humans to flourish in harmony with nature and each other.

As a combination of feminist and ecological concerns, ecofeminism considers anthropocentricism and androcentrism as going hand in hand. The rise of critical animal studies draws attention to intersections between animal cruelty and other forms of oppression to raise awareness to sufferings and operating power politics behind human's treatment of animals. These critiques have proved highly productive for gender, social, and race studies. In feminist literary studies, inquiries into animal metaphors and analogies offer insights and criticism to human history's and society's understanding of and relation to animals. New inclusions of queer studies and new materialism also provide alternative explorations of bodies and identities under oppression and/or seeking transgression. The following three chapters contribute to this growing concern for animals in ecofeminist movements. Anja Höing discusses talking animal stories, Stephanie Baran analyzes animal protection campaigns, and Anja Koletnik attempts to further vegetarian ecofeminist arguments on animal rights.

Höing's chapter, "A retreat on the 'river bank': perpetuating patriarchal myths in animal stores," draws up various talking animal stories from the 1970s to the early 21st century which problematically perpetuate patriarchal ideologies as natural myths. Höing argues that most of these stories operate within a Cartesian dualist framework by gendering talking animal protagonists as male while reducing female existence to passive objects of biological necessity. These patriarchal naturalizations reinforce oppressions of essentialist dualism and master narratives as they deprive females of their agency and materiality.

Turning to magazines, Baran's chapter raises audience consciousness regarding the problem of campaigning animal protection through sexualized and racialized images. Focusing on advertisements released by the People for the Ethical Treatment of Animals (PETA), Baran criticizes the PETA for endorsing a "visualized patriarchy" which abuses rather than aids animals and women along with other marginalized populations. Many of these ads are misogynistic, racialist, and violent, and thus fail to realize environmental justice due to their complicity to the "sex sells" mantra dominating capitalist society.

In "Ethical transfeminism: transgender individuals' narratives as contributions to ethics of vegetarian ecofeminisms," Koletnik proposes "ethical transfeminism" as a means to unbind ecofeminist discussions from dichotomous analytical frameworks. In an attempt to further the study of vegetarian ecofeminism, Koletnik's concept builds off of Carol Adam's and Greta Gaard's canonical studies on meat eating and animal suffering but seeks to expand discussions beyond the limiting usages of gender identity politics. In combing transgender studies with new materialist concepts, Koletnik's ethical transfeminism practices transgressing binary systems in its highlight on ethical self-reflexivity as well as nature and culture/matter bi-directional causalities.

Adriana Teodorescu's chapter, "The women–nature connection as a key element in the social construction of Western contemporary motherhood," also takes to nuancing and challenging models of binary thinking. Teodorescu's chapter observes the contemporary Western construction of the "good mother" paradigm, which idealizes women's connection with nature in an overtly positive and post-evolutionistic manner. She argues that the current stress on the naturalness of childbirth, breastfeeding, and motherhood in fact limits women's choices and disempowers them from their capacity to reason. Having children remains an essentialist value which needs to be deconstructed, first of all, by tearing down the connection between women and a mythologized and perfect nature.

While Teodorescu's chapter criticizes the social construction of women–nature relationships, the next chapter by Denise Mitten and Chiara D'Amore's turns to examining physical interactions between women and nature. Their chapter discusses the many benefits of women–nature contact, in particular how participating in outdoor activities can increase self-concept and relieve women from body image pressures prescribed by society and culture. Their own survey study supports the growing body of data which stress the physical and spiritual productiveness of being in active contact with nature.

Also focusing on women's direct contact with nature, Valerie Padilla Carroll's chapter addresses the potentialities and problems of women's participation in the U.S. back-to-the-land movement during the 1970s. Padilla Carroll focuses on the feminist back-to-the-land magazine, *Country Women*, to illustrate her arguments. For nearly a decade, *Country Women* provided a forum for active feminists to promote self-sufficiency and gender justice. The publication's emphasis on embracing heterogeneity and other ethnicities and species, however, was overshadowed by its fundamental prioritization of the white Western subject as well as existing tensions between desires for individual autonomy and feminist collective sisterhood. In the end, the author argues, these self-writings generated progress but failed to bring about radical change.

The remaining four chapters of this volume expand the scope of environmental protection propaganda and activism discussions across cultural and national borders. Shlomit Tamari's chapter provides a study of female environmentalist

Bilha Givon, who is a well-known personality involved in the environmental protection movements in Israel. To examine Givon's challenging role as a mediator, Tamari brings in Jean-Paul Sartre's theories on "reciprocity" and the dialectics of the "third party" to lay ground for her reading of Givon's interviews, writings, and work. Christina Holmes's chapter investigates various examples of deep intersubjectivity at the Women's Intercultural Center (WInC) near New Mexico's border with Mexico. Employing theory to action, Holmes argues for WInC's facilitation of ecological consciousness through its practice of intersectional feminist environmentalism and celebration of the interconnectedness of life.

The closing two chapters concern environmental issues from India. R. Sreejith Varma and Swarnalatha Rangarajan discuss the life-narratives of C. K. Janu, Leelakumari Amma, and Mayilamma from Kerala. They read these texts as "ecosickness narratives" (Heather Houser 2014) which reveal the "partnership ethics" (Carolyn Merchant 1995) between economic systems, people, and the environment. They argue for the significance of "emotions" in promoting environmental ethics and political actions to create a "human-sensitive environmentalism." The three grassroots ecotexts they discuss advocate a kind of "Earth Democracy" (Vandana Shiva 2006) and optimism for active change. The final chapter, written by Reena Dube, explores the intersection between ecofeminist endeavors and documentary filming in a postcolonial Indian context. Taking the Narmada Bachao Andolan movement as an example of ecofeminist struggle, Dube offers a close reading of Aradhana Seth's *Dam/Age* (2003), which features the postcolonial writer turned transnational celebrity woman activist Arundhati Roy. Dube's analysis details the documentary's innovative and instructive negotiation with power relations between center and periphery as well as art and activism.

The essays collected in this volume aim to forward ecofeminism's move beyond dualism by detailing the intricacies and complications of woman–nature relationships in a critical as well as compassionate manner. As natural beings, commitment to nature seems only natural, but after dualism, how women are to reconnect with nature remains an ongoing challenge and prospect. It is an open question.

References

Adams, Carol J. 1990. *The Sexual Politics of Meat: A Feminist-Vegetarian Critical Theory.* New York: Continuum International.

——. 1995. *Animals and Women: Feminist Theoretical Explanations*, ed. Carol J. Adams and Josephine Donovan. Durham: Duke University Press.

Adams, Carol J. and Lori Gruen, eds. 2014. *Ecofeminism: Feminist Intersections with Other Animals and the Earth.* New York: Bloomsbury Academic.

Alaimo, Stacy. 2000. *Undomesticated Ground: Recasting Nature as Feminist Space.* Ithaca: Cornell University Press.

——. 2010. *Bodily Natures: Science, Environment, and the Material Self.* Bloomington: Indiana University Press.

Alaimo, Stacy and Susan Heckman, eds. 2008. *Material Feminism.* Bloomington: Indiana University Press.

Barad, Karen. 2007. *Meeting the Universe Halfway: Quantum Physics and the Entanglement of Matter and Meaning*. Durham: Duke University Press.
Beyer, Jeff. 2014. "Climate Chaos, Ecopsychology, and the Maturing Human Being." In *Ecopsychology, Phenomenology, and the Environment*, ed. D. A. Vakoch and F. Castrillón. New York: Springer.
Birke, Lynda. 1994. *Feminism, Animals, Science: The Naming of the Shrew*. Buckinghamshire, UK: Open University Press.
Braidotti, Rosi. 2014. "Feminist Philosophy: Coming of Age." In *After Poststructuralism: Transitions and Transformations*, ed. R. Braidotti. New York: Routledge.
Buechler, Stephanie and Anne-Marie Hanson, eds. 2015. *A Political Ecology of Women, Water, and Global Environmental Change*. New York: Routledge.
Cook, Barbara. 2008. *Women Writing Nature: A Feminist View*. Lanham: Lexington Books.
DeKoven, Marianne and Michael Lundblad, eds. 2012. *Species Matters: Humane Advocacy and Cultural Theory*. New York: Columbia University Press.
Estok, Simon C. and Won-Chung Kim. 2013. *East Asian Ecocriticisms: A Critical Reader*. New York: Palgrave Macmillan.
Fishers, Elizabeth. 1979. *Woman's Creation: Sexual Evolution and the Shaping of Society*. Norwell, MA: Anchor Press.
French, Marilyn. 1985. *Beyond Power: On Women, Men, and Morals*. Minnetonka, MN: Olympic Marketing Corp.
Gaard, Greta, ed. 1993. *Ecofeminism: Women, Animals, Nature*. Philadelphia: Temple University Press.
Gaard, Greta. 2001. "Women, Water, Energy: An Ecofeminist Approach," *Organization & Environment* 14(2): 157–172.
Gaard, Greta, Simon C. Estok, and Serpil Oppermann, eds. 2013. *International Perspectives in Feminist Ecocriticism*. New York: Routledge.
Griffin, Susan. 1978. *Women and Nature: The Roaring inside Her*. New York: Harper & Row.
O'Brien, Rory and Matthew Cahn. 1996. "Thinking about the Environment: What's Theory Got to Do with It." In *Thinking about the Environment: Readings on Politics, Property, and the Physical World*, ed. R. O'Brien and M. Cahn. New York: M. E. Sharpe.
Oppermann, Serpil. 2013. "Feminist Ecocriticism: A Posthumanist Direction in Ecocritical Trajectory." In *International Perspectives in Feminist Ecocriticism*, ed. G. Gaard, S. C. Estok, and S. Oppermann. New York: Routledge.
Phillips, Mary and Nick Rumens, eds. 2015. *Contemporary Perspectives on Ecofeminism*. New York: Routledge.
Plumwood, Val. 1993. *Feminism and the Mastery of Nature*. London: Routledge.
Sandilands, Catriona. 1999. *The Good-Natured Feminist: Ecofeminism and the Quest for Democracy*. Minneapolis: University of Minneapolis Press.
Shiva, Vandana. 1988. *Staying Alive: Women, Ecology and Development*. London: Zed Books.
Sturgeon, Noël. 1997. *Ecofeminist Natures: Race, Gender, Feminist Theory and Political Action*. New York: Routledge.
Taylor, Nik and Richard Twine, eds. 2014. *The Rise of Critical Animal Studies: From the Margins to the Centre*. New York: Routledge.
Warren, Karen. 1997. *Ecofeminism: Women, Culture, Nature*. Bloomington: Indiana University Press.

1 Françoise d'Eaubonne and ecofeminism: rediscovering the link between women and nature

Luca Valera

It is often said that the exploitation of Women by Men is a reproachable behavior. Nonetheless, very often we become aware of our destructive capacity with regard to the environment and the exasperated exploitation we are perpetrating with regard to natural resources, thanks, in general, to the press and mass media. These are two different kinds of awareness; they are different in appearance, but they are connected by a unique matrix, that of dominion, which Françoise d'Eaubonne (1974, 220) had the courage and lucidity to bring to light in an explicit manner for the first time in her well-known text *Le féminisme ou la mort*. More specifically, it is the dominion over Women and the dominion over Nature, or to put it better, the dominion of Man over feminine and natural fragility.[1] At the beginning of the 1970s d'Eaubonne condensed in the term "ecofeminism" the idea that it would have been necessary to create a movement of public opinion in order to rescue both Nature and Women. Her intuition was not integrally innovative: the similarity of the condition of Women and Nature (both denominated Mother) was a theme already present in Greek mythology, for which the Mother par excellence is Gaia (or Gea, the Mother Earth), who is generated and auto-generated as the immortal daughter of Chaos:

> In truth, first of all Chasm came to be, and then broad-breasted Earth, the ever immovable seat of all the immortals who possess snowy Olympus' peak and murky Tartarus in the depths of the broad-pathed earth, and Eros, who is the most beautiful among the immortal gods, the limb-melter-he overpowers the mind and the thoughtful counsel of all the gods and of all human beings in their breasts. Earth first of all bore starry Sky, equal to herself, to cover her on every side, so that she would be the ever immovable seat for the blessed gods; and she bore the high mountains, the graceful haunts of the goddesses, Nymphs who dwell on the wooded mountains.
>
> (Hesiod 2006, 13)

The epic of *Gaia*, born beneath the precious mythical clothes told of by Hesiod, will reappear later under the form of scientific theories; for example, in the case of the hypothesis of the British scientist James Lovelock (Lovelock and

Margulis 1974; Lovelock 1986), who would constitute the theoretic background to all ecologisms (and therefore also to ecofeminism).

In this way, ecofeminism inherits that common sentiment that antedates its genesis, just as ecologist ideas precede Haeckel's formulation (Acot 1988) and feminist ideas are prior to the birth of the aforementioned movement. In fact, the thought and historical ground in which ecofeminism blossoms is the French feminist tradition of the second half of the 20th century; in the middle of the 20th century, Simone de Beauvoir underlined that, in the logic of patriarchy, Women and Nature were connected as they present themselves as the 'Other' in respect to the male:

> It is male activity that in creating values has made of existence itself a value; this activity has prevailed over the confused forces of life; it has subdued Nature and Woman. We must now see how this situation has been perpetuated and how it has evolved through the ages. What place has humanity made for this portion of itself which, while included within it, is defined as the Other? What rights have been conceded to it? How have men defined it?
>
> (de Beauvoir 1956, 98)

In the same way, Luce Irigaray brought to light in 1974 the phallocentric logic that constitutes the theoretical background of all identification of Women as strangers to the male universe (Irigaray 1987).[2] If this is truly the historical background of the ecofeminist tradition, it seems that it is the feminist element that prevails over the ecologist one. Ecofeminism, thus, is not characterized as "a special school of social ecology," for it "addresses the basic dynamics of social domination within the context of patriarchy" (Capra 1996, 9). Rather, it is a sort of feminism that is particularly engaged in protecting Nature. As a consequence, "ecofeminists see female experiential knowledge as a major source for an ecological vision of reality" (Capra 1996, 9): the ecologist approach is not sufficient to sustain a feminist revolution.

In order to defend ecofeminism, we could affirm that it "represents the union of the radical ecology movement, or what has been called 'deep ecology,' and feminism" (Ruether 1996, 322). If the true approach of ecology helps feminism to re-contextualize the human being within his own natural environment, healing the fracture between *Homo sapiens* and Nature, then feminism enriches the ecologist prospective through a more aware and complete consideration of the human being.

It is thanks to d'Eaubonne that the most vehement criticism of the distancing of Man from Nature is back in vogue. The critique concerns his presumption of omnipotence generated by technological power and his 'obsession with domination and control' of every living being that is considered inferior (Merchant 1980). Thus, we can say that the tradition of thought inaugurated by d'Eaubonne deals with the patriarchal domination of Women and Nature by Men considered as a paradigm of any domination and exploitation (hierarchical, military, capitalist, industrial, etc.), and with the clear aim of redeeming these two fragile realities from any type of male subjection.

Roots of an affinity: Women and Nature

The core of ecofeminist philosophy is therefore to make explicit the affinity between Women and Nature, between the feminine and the natural universe. In this sense, d'Eaubonne claims that "the relations between the sexes" can be defined as "humanistic and ecological issues at the same time" (d'Eaubonne 1974, 242–243). The first point of similarity between Women and Nature is arguably the concept of maternity: both are mothers. This conceptual issue, which allows us to access the anthropological level of ecofeminism, was certainly the subject of many studies within the feminist universe and has given rise to different strands of thought (Diotima 2007); here, I would like to emphasize the preeminent role that is attributed to the woman as an "authoritative and primary source of both material and symbolic existence" (Cavarero and Restaino 2002, 99) of the child, and the analogous task that is entrusted to Nature.

The most important element of commonality is definitely the analogy of the structural dependence of the child upon the mother (and therefore upon Women), with that of humankind upon Nature, the Mother par excellence: "We are all closely tied to the environment, that our very existence derive from and depends on a healthy environment, as our existence once depended on a mother (or mother-figures, almost always female)" (Roach 1991, 47). Dependence is the first element of similarity, which will introduce the dynamic of exploitation. Here, then, lies the connection between Women and Nature: both are mothers; for this reason, perhaps, "the way we think about and treat the environment is related to the way we think about and treat women" (Roach 1991, 47). If we consider that ecology is the study of the house (*oikos-logos*), the relationship with the activity of women becomes more evident: "It is beginning to dawn on women that they must assume the responsibility for housekeeping nature" (Peterson and Merchant 1986, 465).

Nevertheless, this link presents a problem of great importance: the equation of the dependence of human beings on Nature with that of the child on the mother could, on the one hand, positively remember the debt and the gratitude of the human being toward Mother Earth, and on the other hand, it

> could have the exact opposite effect. Mother in patriarchal culture is she who provides all of our sustenance and who makes disappear all of our waste products, she who satisfies all of our wants and needs endlessly and without any cost to us. Mother is she who loves us and will take care of us no matter what. The last thing the environmental movement should do is encourage us to think of the environment in these terms.
>
> (Roach 1991, 49)

The assignment of a *gender* to the Earth is, therefore, a reckless operation, because, while it succeeds in enhancing the feminine dimension, at the same time it overloads this responsibility, allowing for the possibility of exploitation by males. Moreover, as Teodorescu correctly points out, woman's cultural perception as mother is still stereotypical:

> More than ever, motherhood is a value in itself for contemporary Western society, not only in what concerns its socio-economic importance (providing labor force and national prevail), but also in what concerns woman's cultural perception as mother. Motherhood tends to be viewed as a necessary stage in woman's life which may be subject to delay but which should not be a non-choice, no matter if it completes other dimensions of women's personality or if it represents the ultimate accomplishment (Jong 2010). Popular culture praises motherhood as a stereotypical, sugary display of affection towards an angel-like child through various means—films, news articles, women's magazines, books about child raising.
>
> (Teodorescu Chapter 5, this volume, 00)

The core of Ecofeminist speculation tends to emphasize the affinity between the feminine and the natural universe (defined as everything that man has not modified) and to stigmatize the selfish and utilitarian behavior of men. The definition of Ecofeminism given by one of the protagonists of the movement, Karen J. Warren, is therefore telling:

> As I see it the term eco-feminism is a position based on the following claims: (i) there are important connections between the oppression of women and the oppression of nature; (ii) understanding the nature of these connections is necessary to any adequate understanding of the oppression of women and the oppression of nature; (iii) feminist theory and practice must include an ecological perspective; and (iv) solution to ecological problems must include a feminist perspective.
>
> (Warren 1987, 4–5)

The common fate of 'oppression' of Women and Nature is inscribed from the beginning in their common essence of being mothers; the generation of the child coincides with the condemnation to the child's betrayal: "The child, every child, lives and feeds on the mother's sacrifice: the sacrifice of her time, her body, her space, her sleep, her relations, her work, her career, her affections, and also loves, other than the love for her son" (Galimberti 2009, 17).

The debt of dependency on the mother is often or almost all the time repaid by the child with an even bigger debt: the abuse or the indifference. The reasons for this abuse would be only grounded on *gender* and would have encouraged man to claim the right and power to subjugate the Other.

In addition to this dependence, the link between Nature and Women is made explicit in the concept of care: it attracts and leads immediately to the idea of a complex network (*web*) of relations, which are open to the other living beings, both human and non-human; the network is established by all the beings that surround us and is identified to some extent with the ecosystem: here, the inseparable link between ecology and feminism originates, giving rise to ecofeminism. It is at this level that we find the origin of ecofeminism in the inseparable link between ecology and feminism. The essential connection between Gaia and Women is fully realized, ultimately, in the dimension of care. That issue, however, requires further investigation: it seems that the relation of care, if established with non-human beings, should always be considered one-sided, since only a rational being can be conscious of the possibility of realization of the Other and of the self. It seems impossible then that care could exist at the level of an inter-specific network. The corollary of this statement is the recognition of the fact that Mother Earth cannot take care of her children if not in an 'analogical' manner, and any comparison between Gaia and Women is even weaker. The care given is *between* humans or *by* humans.

If, then, we can conclude that "the central thesis of Ecofeminism is the connection between the social subjugation over women and the domination over Nature [. . .] the critique of anthropocentrism and that of androcentrism must go hand in hand" (Marcos 2001, 148).[3]

The anthropology highlighted by ecofeminism (a connection exists between the mother and Mother Nature) comes to the ethical dimension, by means of the formulation (and the consequent disapproval) of androcentrism.

Indeed, it seems that ecofeminism has a different stance from environmentalism regarding the position of Man in the cosmos: the movement has not the aim to lower human beings as such in the scale of beings, or to raise the other non-human beings, but to annihilate the logic of domination that embodies male. This was the message found in Rachel Carson's *Silent Spring* (1962): Man and Nature are not opposed but are one reality, and, for this reason, the destructive and domineering attitude of men cannot bring any fruit. The point is not to remove the man from the moral summit of the universe because he is a *human being*, but rather to place him within the natural context explicitly because he is a *male*.

The ecological root of *eco*-feminism, then, tends to emphasize the exclusively natural dimension of Man, while, on the other hand, the *feminist* root aims to restore the equalities between the sexes within the human species. In this regard, ecofeminism invites us to overcome Deep Ecology,[4] at least with respect to an essential element: "According to ecofeminists, deep ecologists make the mistake of fighting 'anthropocentrism in general'. What is in question is not the Western world's 'human centeredness', but its 'male centeredness'" (Ferry 1995, 117).

The Copernican revolution proposed here is addressed against the male, guilty of progressively establishing the logic of domination in the course of history, thanks to the 'struggle for survival.' This logic of domination, characteristic of patriarchal societies, evidently brings with it the logic of exploitation of the living beings considered inferior, first of all Women and Nature. As Luc Ferry has shown, the motivation for such exploitation has a threefold matrix: "The first traces this double oppression to the appearance of dualism, the second to that of mechanistic science, while the third base sits directly on difference, on sexually differentiated personality formation or consciousness" (Ferry 1995, 118). Let us analyze further the first two elements mentioned above,[5] in order to understand more thoroughly the consequences of the ecofeminist speculation.

Causes of exploitation: dualism, mechanism, and sexual difference

The critique of dualism is one of the cornerstones of feminist thought and appears frequently as a criticism of aggressive oppositions. Man's conception of the world, in fact, consists in the dichotomous alternation of elements, useful to create a hierarchy in the world. Unlike this tendency, the feminist one seems to be more inclined to glimpse the commonalities rather than the differences; it is for this reason that Men would be more accustomed to competitiveness and contrast, while Women would be more conciliatory and able to mediate between opposite positions.

For this reason, "Ecofeminism is presented as a form of *contextual thinking, pluralistic and holistic*" (Marcos 2001, 148), or as an attempt to stem the masculine 'dichotomism.' The value of the holistic thinking for feminism not only is immediately coherent with the Gaia hypothesis, but also primarily consists in rejecting a hierarchical view of the world, focusing on the relationships between the entities of the systems rather than emphasizing the importance and diversity of individuals and their supposed position in the scale of being. Gaia is not a hierarchy, as it consists of networks, all of which are positioned at the same ontological level.

In the holistic thought, every living being deserves the same ethical respect since it occupies the same position within the system: humans and non-humans are equally important. This fact explains the profound openness to animal life that characterizes the ecofeminist thought: if animals deserve as much dignity as human beings and plants do, it is not clear why rights should be granted only to humans.

Within this holistic 'non-hierarchical' context, there are no dichotomous alternations such as: animated/inanimate, vegetative/animal, non-sentient/sentient, human/non-human, male/female, rational/irrational, body/mind, etc.;[6] on the contrary, holism emphasizes the importance of the whole and the interdependence of its parts. In the whole, in fact, there are not differences, as there is variety and richness. Here, then, ecofeminism once again tends to combine elements of feminism (the adversity to dichotomies) and ecologism (the holistic vision of all), creating a more complete picture of reality and thus facing modern mechanistic reductionism (Warren 1996, xi).

Baconian method has reduced the Great Living Mother into inert matter. Not only Bacon, however: the modern Scientific Revolution—from Descartes to Galileo to Newton—instituted without doubt the basis of the next techno-scientific development, has also reduced the universe to a governable machine, once and for all separating all the world of thought from the world of extension:

> Exterior reality, under the title of *res extensa* entirely detached from the interior reality of thought, henceforth constituted a self-contained field for the universal application of mathematical and mechanical analysis: the very idea of "object" was transformed by the dualistic expurgation.
>
> (Jonas 2001, 35)

The profound unity of the human species with other living beings is, thus, definitively lost in the Scientific Revolution. As Carolyn Merchant writes in *The Death of Nature*: "The world we have lost was organic" (Merchant 2001, 274).

The reduction of the *res extensa* to a mechanical matter has meant, moreover, that the human body itself has been reduced to 'matter probed by scientific instruments' and to an exhaustible resource. The seed of indeterminacy present in an organicist vision is completely eradicated from the idea that every little section of reality can respond to a logic of cause and effect and that, ultimately, can be technologically manipulated. Here we can hear echoes of the Baconian mottos.

The decline of the organicist vision of Nature leaves room in the modern age to a 'lifeless' mechanistic view, crucial to the rise of patriarchal society: the logic of domination and power needs, in fact, a hierarchical view of the living beings. Descartes's thought offers, on the other hand, the suitable aid to endorse the triumph of the *cogito* over the extended world, of the rational over the irrational, imposing such a hierarchical view and giving mastery to Man. The monopoly of the rational knowledge of Nature, brought into being by the objectification of inert reality, can be extended by analogy to the woman, who is the bearer of the emotional seed.

However, it would be inappropriate to include also the male within the mechanistic picture: being a rational animal, he could avoid a reduction of his status to mere givenness. We must also underline the profound difference that emerges from the comparison between the human sexual bodies: the creative activity of man, made explicit in the active force of the sperm, is in contrast to the woman's receptive passivity. This would be another element of commonality between Women and Nature.[7] This consideration reinforces the idea of a supposed superiority of man, allowed to explore and shape inferior bodies for utilitarian purposes; in this regard, Fritjof Capra writes:

> Under patriarchy the benign image of nature changed into one of passivity, whereas the view of nature as wild and dangerous gave rise to the idea that she was to be dominated by man. At the same time women were portrayed as passive and subservient to men. With the rise of Newtonian science,

> finally, nature became a mechanical system that could be manipulated and exploited, together with the manipulation and exploitation of women.
> (1983, 40)

Through the subjection of Nature, Man establishes himself as a 'creator' of the artificial life, which is the summit of culture, as Bacon writes: "In artificial things nature is held in subjugation by the empire of man, for without man these things would never have been made. But through the effort and agency of man we see bodies in an entirely new guise and as a kind of alternative universe or theatre of things" (1996, 455).

The Kingdom of Nature becomes the Kingdom of Man. The Kingdom of the Artificial Life—which is the Kingdom of Man—is profoundly different from the Kingdom of Nature: it is a function of Man himself, since it manifests dynamics that Man is able to control (at least in part). Once Woman has been reduced to Nature (and, therefore, to a resource, to a mere function), she becomes completely controllable and can be subjugated.

In these reflections we find the ecofeminist critique of the artificial reproductive techniques guilty of being a symptom of the reduction of Women to mere "sexual animals,"[8] slaves of their function. Here lies the triumph of the patriarchal societies governed by the male power of science.

D'Eaubonne, the subjugation of Nature and Women, and the 'practical' needs that support ecofeminism

In the discussion of the most relevant issues concerning ecofeminism, we have gone well beyond Françoise d'Eaubonne's philosophical speculation: the main objective of the French activist was, however, to highlight the responsibility of men in the subjection of Women and Nature:

> Practically, the whole world now knows that the two most urgent threats of death are overpopulation and overconsumption; instead, we are a little less aware of the entire Male System's responsibility—precisely because it is male (and not capitalist or socialist)—in these two dangers, but yet very few have discovered that both threats are the logical culmination of one of the two parallel discoveries that gave power to men in the last centuries: their ability to inseminate the Earth like Women, and their contribution in the act of reproduction. Until then, only Women had the monopoly of agriculture and the male believed that the Earth was fertilized by the Gods. At the same time, from the moment he discovered the farming and reproductive possibilities, the "great revolution" occurred for the benefit of Men, as described by Lederer. Once the Earth was taken hostage, and the same happened for fertility (and, therefore, for industry), and for the womb of the woman (and, therefore, for fecundity), it was logical that the exploitation of both would lead to this analogous double danger:

> overpopulation, i.e., an excess of births, and the destruction of the environment, i.e., overconsumption.
>
> (d'Eaubonne 1974, 220–221)

The main reasons that support the emergence of ecofeminism are, therefore, historical reasons, and makes sense only within a practical horizon—as shown by Karen Green:

> feminism and ecology are then linked, not logically or conceptually, but practically, for, when women are not forced to reproduce in order to eat, and when they are given the opportunity to fashion the world that their children will inherit along rational principles, we will be well on the way toward solving the demographic aspects of the environmental crisis.
>
> (1994, 133)

Thus, ecofeminism has a historical and a practical justification (just as its meaning and *raison d'être*[9] are mainly practical): this appears to be a satisfactory thesis, at least from an analysis of d'Eaubonne's writings. The main practical reason of ecofeminism is the following: it is necessary to eradicate all forms of patriarchy, in order to free both Women and Nature from slavery.[10]

On the historical level, however, the establishment of patriarchy coincided with the rise of capital—as a fundamental value of society: "Capital is but the last stage of patriarchy, just like profit is but the last mask of power" (d'Eaubonne 1999, 180).[11] Thus, the patriarchal system is based on the logic of the appropriation of the capital in the forms in which it becomes available: with regard to Women, this results in the appropriation of the reproduction and fertility; and, with regard to Nature, it is expressed in the possible and indefinite consumption of resources. In this sense, the "suppression of patriarchy is not only women's liberation, but hope of salvation for the whole species" (d'Eaubonne 2000, 176). The roots of such thinking can be found, as noted earlier:

> in spite of the fecund powers that pervade her, man remains woman's master as he is the master of the fertile earth; she is fated to be subjected, owned, exploited like the Nature whose magical fertility she embodies. The prestige she enjoys in men's eyes is bestowed by them; they kneel before the Other, they worship the Goddess Mother.
>
> (de Beauvoir 1956, 98)

Ecological and feminist liberation must therefore go hand in hand,[12] because there are many structural, historical, and, above all, practical similarities shared by these schools of thought: "Women are more involved in the ecological problem than men. [. . .] They are more in touch with life and preservation of life. Secondly, the problem of demography involves their freedom since the confiscation of the control of demography means subjection, even slavery" (d'Eaubonne 2000, 176).

If ecofeminism is grounded on a practical need, it will be necessary to identify operational procedures to ensure that this objective (i.e., the liberation from patriarchy, under the forms of overconsumption and overpopulation) can be achieved. In this regard, "this change cannot happen without women being encouraged to play roles in society beyond those of wife and mother. It cannot happen without women receiving education, training, and satisfying careers" (Green 1994, 132). Achieving the ecofeminist goal means, ultimately, undertaking a political and social revolution.

A non-dualistic thought: beyond d'Eaubonne

The revolution undertaken by Françoise d'Eaubonne has a great importance in the history of Western ethics and politics, and it still awaits to be fully recognized. The clear identification of a common logic underlying the two major problems facing contemporary societies (the ecological crisis and the denial of the equal dignity of Men and Women) has allowed us to find some possible political and educative solutions.[13] The effort undertaken by d'Eaubonne—carried out over the years by numerous thinkers and activists of the ecofeminist movement, such as Val Plumwood, Karen J. Warren, Maria Mies, Vandana Shiva, and Ariel Salleh—has the great value of having brought to a synthesis, within a single thought, the intellectual adventures of different philosophical traditions. Thanks to d'Eaubonne, we can say today that the expropriation of the female body and Nature fall under a single dominant approach, which has led to almost irreparable damages.

This dominant attitude is reflected, at the ethical level, by a rationalist attempt to codify the laws that govern human action, without considering the importance of the emotional insight and the peculiarities and the "genuine virtues" (Green 1993, 386) through which males and females interpret the world of experience; at the political level, the ethical domain is expressed in the rise of utilitaristic capitalism, which tends to regard the other as a resource only. For both problems, d'Eaubonne reacts with a revolution based on a well-codified anthropology:

> Capital [. . .] will only disappear with an ecological solution of production (and of consumption) which will now constitute the only possible elimination of the outdated structures of dominance, aggressiveness, competitiveness, and absolutism in order to replace them with those of cooperation and equality between individuals (thus between sexes), and of the species with the environment.
>
> (d'Eaubonne 1999, 181)

We should also be grateful to ecofeminism for regarding man as a part of nature, after years of speculation on his mere eccentricity (Plessner 1928). The Man of d'Eaubonne—and, with him, the Women—is not an "incarnated spirit,"[14] but a corporeal being. The human being should be grateful to nature both for his body and instincts and for his rationality and freedom. Furthermore: being 'a rational being' is the greatest form of gratitude manifested to Nature by human beings, as this is their nature.

In this regard, the most significant attempt made by d'Eaubonne consists in overcoming the dichotomies—both anthropological and cosmological—that too often have characterized Western thought. Yet, the attempt does not appear entirely successful, as Karen Green rightly notes:

> Nevertheless, in their own way, they have remained caught within these dualisms by suggesting that because women are not inferior, men must be, particularly in their psychosexual make up, and by accepting the ideas that nature is more important than culture, emotion a better guide than reason, the body more important than the mind. None of these pairs involve opposites.
>
> (1994, 133)

It seems that the positive and proactive aspects which ecofeminism covers are likely to succumb to the dichotomous logic that has been so much criticized: the same instrument that is put under investigation (criticism) is reused for the *pars destruens* of the patriarchal system and for the *pars construens*. Thus, ecofeminism seems to abuse the same instrument that it seeks to criticize, endorsing an 'anthropocentric' anthropology and ethics, which is likely to disqualify the achievements of such a richly intellectual movement.

The recognition of the difference must be—and here lies the interesting reflection which can be developed to enhance d'Eaubonne's proposal toward a non-dualistic anthropology (and ethics)—a motivation to recognize the unity of the human being. Within this unity, rationality and instinct, feelings and logic, fruitfully co-exist: starting from this acknowledgment we can imagine an ethic that is nurtured by the contribution of feelings and passions (and that is not characterized exclusively as a rationalist analysis of our actions or as a mere risk/benefit calculation), and which aims at a human flourishing in harmony with Nature. In this regard, the ethics of care[15] that feminism and ecofeminism gave to our society appears to be the best effort to positively consider the complexity of the human being and Nature of which humans are a significant part.

Notes

1 On this topic Karen Green (1994, 121) wrote: "The case against the subordination of women extends to a case against all the relation of subordination; in particular, it implies a case against the subordination of nature. I call this position the first logical argument for ecofeminism: *feminism implies deep ecology*."

2 For a more precise definition of the concept of 'Otherness' by Irigaray and the above mentioned de Beauvoir, please see Green (2002).

3 This thesis can be read in d'Eaubonne 1999 (180).

4 For a more detailed analysis of the similarities between the two perspectives and of the originalities introduced by ecofeminism, see Salleh (1984). Underlying the connection between deep ecology and ecofeminism, Green (1994, 122) writes: "Often ecofeminists

make the slightly weaker claim that the adoption of feminist values and a gynocentric world view is necessary in order for a new ecological order to be established. Accordingly, only if feminism is adopted will the values implicit in deep ecological thought be accepted. I call this position *the second logical argument for ecofeminism: deep ecology requires (hence implies) feminism.* To defend this position, one needs to show that by adopting the deep ecologist's claim that there are values in nature that are not anthropocentric, we are adopting values that are gynocentric."

5 We will focus exclusively on the first two elements, as they seem to be more significant and in need of a more detailed explanation. Regarding the third aspect, however, please refer to Dodson Gray (1979, chs. 4 and 5).

6 Val Plumwood (1995, 156) thus wrote: "As ecofeminism points out, Western thought has given us a strong human-nature dualism that is part of the set of interrelated dualism of mind-body, reason-nature, reason-emotion, masculine-feminine and has important interconnected features with other dualisms. This dualism has been especially stressed in the rationalist tradition."

7 In this regard, De Beauvoir (1956, 175) wrote: "Man expects something other than the assuagement of instinctive cravings from the possession of a woman: she is the privileged object through which he subdues Nature."

8 The picture, certainly strong but effective, is borrowed from Sarti (2006, 208).

9 The demonstration that the most significant *raison d'être* of ecofeminism is a practical one is provided by d'Eaubonne (1974, 221) herself, who dedicates much space in her writings to the promotion of a feminist revolution: "The only mutation to save the world in our times is that of a 'great revolution' of male power that has brought, after the agricultural exploitation, to a mortal industrial expansion. Neither the 'matriarchy', thus, nor the 'power of Women', but the destruction of power *by* Women. And, then, the end of the tunnel: the equal management of the world for a renaissance (and no more for a protection, just as the first ecologist still believe)." See also d'Eaubonne (1977, 26–28).

10 In this regard, d'Eaubonne's (d'Eaubonne 1999, 184) conclusion at a talk in Canada in 1980 is particularly significant: "Only Ecofeminism will put an end to Patriarchy and save human society from the devastation wrought on the environment, the nuclear threat and the profit-based system which is at the origin of all war and exploitation of this planet." On the same topic, see Archambault (1993, 19).

11 In another significant text, d'Eaubonne (1974, 235) underlines the dynamics of male power regarding simultaneously the ecological issues and the relationship between the sexes: "At the base of the ecological problems, there is the structure of a certain power. Just as in the case of overpopulation, it is a Man's problem; not only because it is man who holds the world power [. . .] but also because the power, at an inferior level, is distributed in a way that Men exercise the power over Women."

12 As the founder of ecofeminism writes, "ecological liberation is not only a liberation of Man's exploitation of Nature but also the liberation of ecology itself from the 'dictatorship' of the two main issues: the exhaustion of resources and the environmental devastation: 'Ecology, the science that studies the relationship between living things within the physical environment and their evolution,' comprehends, by definition, the relationship between the sexes and the birth rate that comes forth from it; its interest is orientated in reason of the horrors that threaten us, towards the abuse of the resources and the destruction of the environment. And so we have arrived to the moment to remind certain other elements that closely intercept the female question and her struggle" (d'Eaubonne 1974, 223).

13 In an impressive critical piece on the capitalist system, d'Eaubonne (1999, 184) wrote: "It is impossible, within patriarchy, to suppress a market economy. And it is impossible, in a market system, to not devastate the planet."

14 This expression is taken from Lucas Lucas (1993).

15 In this regard, "our obligations to care for others, whether they are children, animals, species, or ecosystems, are not merely irrational feelings [. . .]. Our moral judgments may

properly be associated with appropriate feelings, so that feeling is by no means irrelevant to ethics" (Green 1994, 133).

References

Acot, P. 1988. *Histoire de l'écologie*. Paris: Presses Universitaires de France.

Archambault, A. 1993. "A Critique of Ecofeminism," *Canadian Woman Studies/Les Cahiers de la femme* 13(3): 19–22.

Bacon, F. 1996. "Parasceue ad historiam naturalem." In *The Oxford Francis Bacon*, ed. G. Rees and L. Jardine. Oxford: Clarendon Press.

Capra, F. 1983. *The Turning Point. Science, Society and the Rising Culture*. New York: Bantam Books.

——. 1996. *The Web of Life. A New Understanding of Living Systems*. New York: Anchor Books.

Carson, R. 1962. *Silent Spring*. Boston: Houghton Mifflin.

Cavarero, A. and F. Restaino, eds. 2002. *Le filosofie femministe. Due secoli di battaglie teoriche e pratiche*. Milano: Bruno Mondadori.

d'Eaubonne, F. 1974. *Le féminisme ou la mort*. Paris: P. Horay.

——. 1977. "La mère indifférente," *Les Cahiers du GRIF* 17–18: 25–28.

——. 1999. "What Could an Ecofeminist Society Be?" *Ethics and the Environment* 4(2): 179–184.

——. 2000. "Feminism–Ecology: Revolution or Mutation?" *Ethics and the Environment* 4(2): 175–177.

de Beauvoir, S. 1956. *The Second Sex*, trans. H. M. Parshley. London: Jonathan Cape.

Diotima. 2007. *L'ombra della madre*. Napoli: Liguori.

Dodson Gray, E. 1979. *Why the Green Nigger? Remything Genesis*. Wellesley: Roundtable Press.

Ferry, L. 1995. *The New Ecological Order*, trans. C. Volk. Chicago: University of Chicago Press.

Galimberti, U. 2009. *I miti del nostro tempo*. Milano: Feltrinelli.

Green. 1993. "Reason and Feeling: Resisting the Dichotomy," *Australasian Journal of Philosophy* 71(4): 385–399.

——. 1994. "Freud, Wollstonecraft, and Ecofeminism: A Defense of Liberal Feminism," *Environmental Ethics* 16(2): 117–134.

——. 2002. "The Other as Another Other," *Hypatia* 17(4): 1–15.

Hesiod. 2006. *Theogony. Works and Days. Testimonia*, trans. G. W. Most. Cambridge, MA, and London: Harvard University Press.

Irigaray, L. 1987. *Speculum of the Other Woman*, trans. G. C. Gill. Ithaca: Cornell University Press.

Jonas, H. 2001. *The Phenomenon of Life. Toward a Philosophical Biology*, trans. L. Vogel. Evanston: Northwestern University Press.

Lovelock, J. E. 1986. "Gaia: The World as a Living Organism," *New Scientist* 18: 25–28.

Lovelock, J. E. and L. Margulis. 1974. "Atmospheric Homeostasis by and for the Biosphere: The Gaia Hypothesis," *Tellus* 26(1–2): 2–10.

Lucas Lucas, R. 1993. *L'uomo spirito incarnato. Compendio di filosofia dell'uomo*. Cinisello Balsamo: Edizioni Paoline.

Marcos, A. 2001. *Ética Ambiental*. Valladolid: Universidad de Valladolid.

Merchant, C. 1980. *The Death of Nature: Women, Ecology, and the Scientific Revolution*. San Francisco: Harper & Row.

——. 2001. "The Death of Nature." In *Environmental Philosophy. From Animal Rights to Radical Ecology*, 3rd edition, ed. M. E. Zimmermann et al. Upper Saddle River: Prentice Hall.

Peterson, A. and C. Merchant. 1986. "Peace with the Earth: Women and the Environmental Movement in Sweden," *Women's Studies International Forum* 9: 465–479.

Plessner, H. 1928. *Die Stufen des Organischen und der Mensch. Einleitung in die philosophische Anthropologie*. Berlin: Walter de Gruyter.

Plumwood, V. 1995. "Nature, Self, and Gender. Feminism, Environmental Philosophy, and the Critique of Rationalism." In *Environmental Ethics*, ed. R. Elliot. Oxford: Oxford University Press.

Roach, C. 1991. "Loving Your Mother: On the Woman–Nature Relation," *Hypatia* 6: 46–59.

Ruether, R. R. 1996. "Ecofeminism: Symbolic and Social Connections of the Oppression of Women and the Domination of Nature." In *This Sacred Earth. Religion, Nature, Environment*, ed. R. S. Gottlieb. New York: Routledge.

Salleh, A. 1984. "Deeper than Deep Ecology: The Eco-Feminist Connection," *Environmental Ethics* 6(4): 335–341.

Sarti, M. A. 2006. "Ecofemminismo e natura." In *Una nuova etica per l'ambiente*, ed. C. Quarta. Bari: Dedalo.

Warren, K. J. 1987. "Feminism and Ecology: Making Connections," *Environmental Ethics* 9(1): 3–20.

——. 1996. *Ecological Feminism: An Overview of the Issues*. Bloomington: Indiana University Press.

Part II
Rethinking animality

Part II

2 A retreat on the "river bank": perpetuating patriarchal myths in animal stories

Anja Höing

Animal stories, especially anthropomorphic ones, are chronically underrepresented in scholarship. Traditional literary criticism disregards them as a minor side-branch of children's fiction in which metaphoric animals idle through Arcadian worlds, teaching lessons on the beauties of nature and the importance of conservation. Stories, particularly from the second wave of talking animal stories in English literature, which started in the 1970s and ebbed at the turn of the century, have rarely been read against another grain other than for their political correctness in the green cause. Yet when interlinking a traditional critical approach with an ecofeminist reading, one can see that these stories deserve far closer attention than they usually get. Using non-human protagonists as a shield, authors can utilize the ambiguity of the 'animal' and the 'natural' as a protective sandbox that allows hiding patriarchal ideologies behind biologistic arguments.

As Baker points out in *Picturing the Beast*, "ours is a culture . . . in which animals can apparently be used to mean anything and everything" (2001, 4). True to this claim, the animal protagonists of talking animal stories are used for a wide variety of purposes. They live model lives of sustainability in unison with an idealized, Arcadian nature, serve as examples of peaceful co-existence of species or preach fervently on pantheism and Gaia worship. Yet, one can read talking animal stories in a different light when using as a lens for the central claim of ecofeminist theory, namely "that the ideologies that authorize injustices based on gender, race and class are related to the ideologies that sanction the exploitation and degradation of the environment" (Sturgeon 1997, 23). I claim that whilst crying out against the exploitation of nature many talking animal stories of the late 20th century strongly reinforce the closely related exploitation of women. By means of gendering the animal as male and patriarchy as 'natural,' such animal stories create male universes in which females are objectified and dangers to the patriarchal status quo are closely associated with dangers to the environment. In consequence, these stories propagate deeply conservative, ultra-patriarchal ideologies and construct these as solutions to the current environmental crisis.

At this point, an aside is necessary. In the following I am going to refer to dominant tendencies in the genre during the last decades of the 20th century, but my claims are by no means universal. There are beautiful exceptions in stories such as Gary Kilworth's *Hunter's Moon*, which presents vixen and fox on eye level, or in Barbara Gowdy's *The White Bone*, which deliberately turns the tables on the pattern sketched above and pointedly reverses traditional gender roles into a matriarchal society.

A final note on terminology: dualisms are so inherent to Western culture that they are deeply entrenched into our linguistic structure and sometimes expression without them becomes overly complicated and in consequence breaks the flow of reading. Therefore, for the sake of both convenience and readability, I will retain the classic dualist and anthropocentric terminology of human vs. animal. I am aware of the problems intrinsic to this dichotomy, yet convinced that none of these problems could be solved by applying currently popular alternatives such as human/non-human or human animal vs. non-human animal, which, in my opinion, rather highlight the opposition they seek to break up.

As Val Plumwood states in her theory of the mastery model, in Western ideological constructions the lot of nature is generally cast with the one of women (1993, 33). Both are contrasted to the 'superior' male realm of culture (43). 'The animal' as an abstract concept normally appears as a part of 'nature.' By inference the animal, though not explicitly mentioned in Plumwood's "key elements in the dualistic structure of western thought" (43) would be placed on the same side of the dichotomal division as the female. Indeed, with regard to the cultural construction of the place of animals in Western modes of thought, one can find ample proof for this connection. Animals generally appear as 'the other.' One only needs to remember Derrida's famous experience of the ultimate other in the gaze of his cat (2002, 380). While Derrida's experience, however, led him to engage with the animal and pose a set of questions to reach an understanding of this other (418), the Cartesian tradition rejects any possibility to bridge to the animal. Cartesian thought treats animals as *automata* governed by a mechanistic, non-deliberate instinct diametrically opposed to human, or, in traditional Western dualist structure, *male* rationality. Additionally Western sciences generally treat animals as governed solely by necessity, or, more precisely, the impulse of survival. Animals are cast as primitive in contrast to civilized man, as concerned with reproduction, not production, and finally as passive objects instead of active subjects. All these aspects clearly place the animal on the same side as the female in Plumwood's mastery model.

Despite this, there is an ongoing tradition of gendering the literary talking animal as male and thus dissociating it from the real-life animal, which, using Baker's terminology, is degraded to "the sign of all that is taken not-very-seriously in contemporary culture" (2001, 174). The literary talking animal in contrast is generally the agent of a story and therefore has to be taken seriously at least by the narrator of its own tale. To enable the implied reader (in general a white, male implied reader) to identify with the animal protagonist, its otherness needs to be partly obliterated, and the most prominent way to do so appears to be gendering the animal as male.

The tradition to do so is probably as old as the talking animal story itself. A well-known precursor is certainly *The Wind in the Willows* (1908). Kenneth Grahame's story is one of the immortal tales of children's fiction: on the Arcadian "river bank" (2008, 1), Mole, Rat, Badger, Toad and their friends enjoy a carefree life full of little adventures and great joys. Although Grahame's tale was by no means the first story presenting an all-animal-society, it set the pace for talking animal stories for a century to come. Even nowadays, many animal stories are advertised as, for example, "shar[ing] *The Wind in the Willows'* charm" or able to "rival *The Wind in the Willows*" (Moorhouse 2013, n.pag.).

Despite the book's lasting success, *The Wind in the Willows* has long been notorious for its all-male world. All animal protagonists are males. Female animals apparently do exist somewhere: the "Otters" (Grahame 2008, 118) are a married couple and the (male) mice children have "small sisters" (95) at home. But these females are entirely invisible and of no consequence to the story. Not unjustly, thus, have critics claimed *The Wind in the Willows* to be "misogynistic" (Cosslett 2006, 151; Hunt 1994, 86) and to carry "male dominance . . . to a rare extreme" (Hunt 1994, 86). As easily as one can detect the "Weasels of the Wild Wood" as the looming threat of the uprising working class (79), one can read the exclusion of women from the animals' happy life as the Edwardian gentlemen's longing for a nostalgically oversimplified world of unthreatened patriarchal power. Grahame himself described his story as "clean of the clash of sex" (qtd. in Hunt 1994, 85). *The Wind in the Willows* belongs, as Manlove contends, to a "threatened conservative culture" (1999, 5). In defence of this culture Grahame constructs a bubble in time in which patriarchy may continue in eternal safety.

The cultural frames of the "river banks" of the late 20th and early 21st centuries might have shifted, yet the basic structure often remains the same. Hardly less conspicuously than in *The Wind in the Willows*, sometimes even more so, the world of talking animals still tends to be a thoroughly male-dominated one. The first thing that strikes the eye when reading almost any talking animal story is a distinct shortage of female characters. In Richard Adams's *Watership Down* (1972), the most famous late 20th-century talking animal story, of the eleven rabbits who set out to seek a new home, not a single one is female and apparently it does not occur to the characters at a single point to induce any females to come with them, although their objective is to collect "as many [rabbits] as [they] can" (Adams 1974, 27). In fact, during the chapters of the book set in the rabbits' home warren, one might be induced to think that there are no females there at all; neither narrator nor characters mention a single one. 'The rabbit,' it seems, is male.

It is not only 'the rabbit' in *Watership Down* that is constructed as male, but also in fact 'the talking animal' as a concept. A prime example would be Colin Dann's *Animals of Farthing Wood* (1979). Dann's animals do not have names, but species designations; it is 'fox' and 'badger' who lead their journey to a nature reserve, and 'hare' and 'rabbit' who follow—a typical device for animal stories featuring multispecies protagonists (Cosslett 2006, 175). All the characters designated as 'species' are male. Comparable ways of constructing the talking animal as male also appear in other, more recent talking animal stories; for example, Peter Chippindale's *Mink!* (1995).

Robin Hawdon's ant story *A Rustle in the Grass* (1984) uses another, more self-aware method of gendering the animal. Every ant of the colony, except for the Queen, is given a male pronoun and a masculine name. This will deeply confuse a reader with any basic biological knowledge, yet discourse apparently demands male heroes, even when these are a biological impossibility. The same issue can be observed in animated pictures featuring ants—both Disney's *A Bug's Life* and Dreamworks's *Antz* use male ants in biologically female-only positions. Interestingly, *A Rustle in the Grass*, as if embarrassed by its own device of changing the ants' gender, offers a plot-intrinsic explanation. The ant protagonists tell stories that their god banned all males from the colonies as a punishment for misbehavior (Hawdon 1984, 226–227) and the females in consequence had to assume all male "functions as rulers, soldiers, builders, hunters and even as ordinary workers" (227) and finally "took their titles away from the males and became masculine even in name" (227). This explanation may arrive at the reader's mind as if, first, the pronoun 'he' was just so much more desirable than 'she'—why else would the females be so eager to change their gender designation?—and, second, that one has to be a 'he' in order to fulfill certain functions in a sophisticated society. To achieve power, to, in fact, become credible when operating in the domains of Western hegemony borders as traditionally male, characters that can only be female in sex have to usurp male gender in discourse.

Defined as male, the animal becomes easier to tackle for authors writing inside Cartesian dualist structures. A male/cultural animal can serve as a subject without the need for the author to think beyond pre-established categories. Like Grahame's Water Rat who becomes physically ill at the sight of new horizons (2008, 172) and has to turn his gaze back to the river bank in a stream of romantic poetry to recover after the frightening ordeal (175) such stories break down the otherness of the 'animal' into a blank slate that can be used as metaphor for the masculine self. As Plumwood states, "in the colonising strategy of the master, only what is marked as Self is permitted to survive" (1993, 195). The strategy of many talking animal stories is accordingly one of annexing the animal into this concept.

As soon as the animal appears as a metaphor for the self, and the self is a male/cultural one, the literary animal's association with 'the natural' becomes problematic. The protagonist, however dissociated from the connotations normally linked to the term, is still labelled 'animal' and accordingly 'natural' by association, while its male gender links it to the 'cultural.' On a first glance, this appears to be a dilemma unsolvable without moving beyond dualist thinking, but many animal stories again solve the problem not by widening the horizon, but by shifting the gaze. In fact, the construction of the animal as male encounters an equally ideologically loaded construction of 'the natural.'

The definition at work when we nonchalantly label an animal as 'natural' is primarily the one that understands 'nature' as "the phenomena of the physical world collectively; *esp.* plants, animals, and other features and products of the earth itself, as opposed to humans and human creations" (*Oxford English Dictionary* "Nature, n."). Even the concept of nature in this material definition is problematic as it perpetuates the dualism between nature and culture which is deeply embedded in Western ideology, even though this dualism is ultimately "self-defeating" (Gruen 2011, 48). Yet, as Gruen rightly states, "the concept of the 'natural' is a fraught one" (46). In everyday discourse as well as in talking animal stories, 'natural' also may refer to "a norm against which deviation is measured" (Morton 2007, 14). When animal protagonists blame another animal for being "unnatural"—by no means an uncommon accusation in talking animal stories—this does not primarily refer to physical but to normative deviations from what is constructed as the right way of life. 'Nature' thus becomes another expression for the hegemonic structures governing the animal societies, while at the same time still being applied to the physical world.

In consequence, many talking animal stories blur the distinguishing line between the material and the normative concept of nature up to the point where the two become indistinguishable. What is materially 'natural' is constructed as normatively 'natural' as well—a naturalistic fallacy that allows an attack on any deviation from the norm as not ideologically and accordingly subjectively problematic, but *materially* so, and thus can be used to justify prescriptions. Ideological structures thus are transformed into essentialist notions. In consequence, biologistic arguments present ideologies and hegemonic structures as "given, inescapable, and therefore moral," using Donna Haraway's definition of the problem (1991, 8). Haraway rightly states that such "differences seen as natural" are a major source of domination (8), as the designation as physically natural renders the differences untouchable by arguments assailing them from the ideological side.

In talking animal stories, a society in accordance with the 'natural order' based on this chimerical concept of nature most often is a local, pre-industrial androcentric society with a deeply conservative attitude, modeled upon idealized and overly simplified conceptions of primitive human societies as living in harmony with nature. The talking animal characters, in certain dimensions, take the place of what Padilla Carroll (2017) describes as the "ecological Indian trope" (Padilla Carroll 2017). Although their societies are more often based on romantic notions of pre-Christian British tribes or medieval communities, the underlying idea of "indigenous oneness with the land" (Padilla Carroll 2017) is almost identical. Adams, for example, directly equates his rabbit characters to "primitive humans" (1974, 169) and the rat protagonists in Robert C. O'Brian's *Mrs. Frisby and the Rats of NIMH* deliberately model their society on a medieval monastery which they present as the perfect model of a closed self-sustainable community (1975, 137). The animals' life is 'natural' in that it takes place in the non-human physical world, but due to the chimerical concept of nature their patriarchal society structures are presented as 'natural' as well.

The—physically—natural system that sets the framework for animal life is the ecosystem, and in fact it is this very concept that the stories try to translate into society structures. But while in other contexts links between women and nature are translated into the trope of the Ecosystem (Valera 2017), here it is patriarchal power structures. Not surprisingly, these ecosystem-societies hardly reflect ecological reality, but rather the myth that an intact ecosystem operates in "achieved and unchanging harmony" (Colleen Clements qtd. in Garrard 2012, 64). This concept of ecosystems completely disregards the biological reality of succession and of flow equilibrium, but nonetheless is the most widely accepted model in Western societies (Garrad 2012, 64). The model cements the myth of nature as stable and unchanging—as, in a word, conservative. In an inversion of the argument that applies the chimerical concept of nature to this misconception of ecological balance, this idea of ecosystems reinforces the myth that conservatism is 'natural.' As conservatism is also patriarchal, 'the natural' operates as a whitewash for ultra-patriarchal animal societies.

The easiest way to keep the 'ecosystem' of animal society in eternal patriarchal stability would be the one used by Kenneth Grahame—to circumvent the "threat" (Hunt 1994, 87) of females by excluding them entirely. Yet, females are not a topic that can be eluded without cost. In order to give any chance of both credibility and continuity to the Arcadian patriarchal retreat of the late 20th century "river bank," the animal cultures need females for the simple reason that they need to produce a new generation of male protagonists. Females, in short, are a biological necessity. Not surprisingly many stories reduce females to precisely this dimension: "sexual animals," as Valera aptly borrows from Sarti, who are "slaves of their function" (Valera 2017). They appear as sexual objects, trophies and possessions. Referring to Brian Jacques's *Redwall* (1986), Holly V. Blackford states that Jacques's mouse hero's "rise to power depends on the patriarchal exchange of women as a sign of sexual power and economic status" (Blackford 2003, 97). The hero is given the female mouse Cornflower "as a reward" (97) for his deeds, much as, as Baran describes, men are promised women as trophies for a vegan lifestyle in PETA advertisement (Baran 2017). Yet, Cornflower is most centrally constructed not as trophy for, but as "*mate* for the hero" (102, italics mine) and as such she has another even more central function which I would add to Blackford's findings: she is there to mother the hero's son, who in turn presents the future of the Arcadian Redwall community (Jacques 2002, 350). As Teodorescu (2017) discusses in the more general context of Western culture's "good mother paradigm", female protagonists in talking animal stories, just like contemporary Western women, are constructed through motherhood. To use a talking animal story narrator's words: they "fulfil their lives by raising cubs" (Clement 1988, 317). In Clement's *The Cold Moons*, and in many other stories as well, reproduction is the female animals' sole function.

A prime example of the streamlining of female characters into the role of mate/mother is Mary Stanton's *The Heavenly Horse from the Outermost West* (1988), thus proving that the pattern cannot entirely be attributed to the gender of the author. In what is unquestioningly a scene of a rape attempt—but is never called one, a reminder of how language constitutes the world—the rebellious female protagonist of the story, Duchess, has to learn the hard way "that stallions have their Rights" (117), "granted by Law" (117). The capital-letter law in question is a divinely ordained "Balance" (10) that subjects Duchess to her biological function as mate to the stallion. Poetic justice follows on foot when she rejects his mating attempts. Duchess finds herself continuously suffering until the moment she happily submits herself to her sole function: she bears the stallion's child, the, of course, male scion to his line (306).

Richard Adams's *Watership Down* shows the ideological foundations behind such patterns. Having finished their journey, the rabbit heroes laze on a sunny hillside, when one is suddenly struck by what has escaped all of them until that moment: that their brand-new warren is "as good as finished" (1974, 195) because they have "no does—not a single one—and no does means no kittens and in a few years no warren" (195). Again, there will be no new generation of male protagonists without females to reproduce them. As in the "good mother paradigm" (Teodorescu 2017), the importance of the mother virtually evaporates behind the one of the child. In talking animal stories she is only a means to an end: the focus is on the continuity of the patriarchal community. Adams appears to be aware that this entirely functional approach to females might be uncomfortable for some readers and adds the following explanation:

> The kind of ideas that have become natural to many male human beings in thinking of females—ideas of protection, fidelity, romantic love and so on—are, of course, unknown to rabbits, although rabbits certainly do form exclusive attachments much more frequently than most people realize. However, they are not romantic and it came naturally to Hazel and Holly to consider the two Nuthanger does simply as breeding stock for the warren. (256)

Here one can see at work what Barthes in his *Mythologies* refers to as "naturaliz[ation]" (1974, 129). *Watership Down* presents romantic protective notions as a culturally constructed myth, yet the patriarchal dualistic opposition between male subjects and female objects—nothing but another cultural myth—is taken for granted. This myth has, to use Barthes's words, been transformed from "history into nature" (129). While the stories naturalize that males are agents, females are naturalized as breeding and mothering machines. This is precisely the same mode of naturalization one can see at work in the PETA advertisements discussed by Baran (2017). Even the ants of Robin Hawdon's *A Rustle in the Grass* (1984) longingly look back to an Arcadian past when "they lived and progressed together, male and female in harmony, the females breeding and caring for the young, and the males doing the hunting and fighting and the heavy work" (226).

While the male animal thus becomes a metaphor for the masculine self, the female is constructed as the Cartesian *automaton*. This is further reinforced by a technique of de-individualization: secondary female characters are often refused a name and referred to only as a group. In a raid on another rabbit warren in *Watership Down*, most does the heroes take with them remain unnamed, while not a single male rabbit does. The same pattern can be observed in *The Cold Moons* (Clement 1988). The badgers are joined by "Gnos and two of the sows" (316) or leave "ten senior female badgers all under the supervision of Harvey" (71) to take care of the young. Leaving the female characters unnamed and referring to them only in plurals deprives them both of their agency and of their identity as characters and in consequence eases treating them as mere attachments to the males.

The same objectification of females is implicit in the possessive denomination of female animals in *The Animals of Farthing Wood*, where they are reduced to their sexual function as the male's "mate." But even denominations such as "Hare's mate" (Dann 2008, 34), which at least grant a clear designation to the female, albeit a possessive one, are rare. Far more frequently the females occur as afterthoughts attached to the male animal, as in "Pheasant and his mate" (9). When the hen pheasant is shot shortly after her mate, the animals do not even comment on her death. The moment they remark that "Pheasant's done for" (99), she appears to drop out of existence—she in fact *does* not exist once she can no longer fulfill her reproductive function. This also explains one of the logical breaks of the story: how vixen, from the moment she meets fox being entirely dependent on his protection, could have survived on her own prior to his appearance. The answer is a simple one: she didn't. She did not exist, cannot, in fact, exist, without being attached to a male.

The protagonist in *Mrs. Frisby and the Rats of NIMH* similarly exists only through the connection to her deceased husband. Unlike all other major protagonists in the story, she has no first name, but only is "Mrs Jonathan Frisby" (O'Brian 1975, 66). Ratelle argues that "for O'Brien . . . this absence [of a given name] is not a site of subjugation but a signification of non-subjectivity that defines the female mouse as the de-individuated nucleus of a trans-species collective" (2015, 109) and concludes that O'Brien "employs traditional gender and species expectations to extend a posthumanist model of subjectivity to all animals" (109). I would argue for a much simpler reading. Everything the rats do for Mrs. Frisby they do in memory of her husband—in fact, they categorically refuse to help until she is brought into connection to the heroic deeds of "Mr Jonathan" (66). Mrs. Frisby, though certainly de-individuated, does not appear as a mash in a net of trans-species subjectivity, but as an object through which the help of the rats is transferred to both the memory and the genes of her husband in form of her son Timothy.

The objectification of females in talking animal stories goes so far that they sometimes drop out of existence altogether while they are not needed for the plot. This is prominent in the final scene of *The Animals of Farthing Wood*. Fox "allow[s] the male animals forward to sample" a puddle of beer they come across (Dann 2008, 300). While they do so, speak various toasts and get drunk, the female spectatorship virtually fades out of the scene. When the story ends a few pages later, it appears as if the "animals [who] unconsciously draw together into one group as the priceless feelings of friendship and loyalty enter[s] again, and for ever, into each heart" (302) were all of them. But they are not: they are the males only. The same occurs in *The Cold Moons.* At the beginning of the story the narrator explains that "all boar badgers over four years old" (Clement 1988, 37) are allowed to vote in the community, but shortly after one of the heroes remarks that "all mature badgers" (47) do so. At this point, sows have dropped out of the concept of 'badger' altogether. In contrast, once females morph into the concept of badger, they lose their sex. When the evil badger Kronos deserts from the main group, he takes male and female badgers (145), but the next time they meet, Kronos's group suffers from "the lack of caring females" (256) and to remedy this they entice some of the main group's females to join them. So what happened to the females who originally went with Kronos? The moment they became part of a group of 'badgers'—a male concept—they dropped out of existence as 'females.'

In addition to their central function as reproduction machines shelved when not needed, female animal characters often appear as the traditional 'angel in the house.' Many female characters are "shy" (Dann 2008, 172), "gentle" (Clement 1988, 89) and "kind" (Horwood 1980, 49), looking up in "admiration" (Clement 1988, 87) to their "gallant" (Dann 2008, 172) mates. Often they are entirely domestic and blissfully content in this role. As Blackford pointedly remarks, the job of Cornflower in *Redwall* is "to serve the food" (2003, 102)—and this is it, even while the community is under siege and every hand, or rather paw, would be needed to fend of the attack. Similarly, in *Watership Down*, none of the female rabbits ever joins in the adventures of the males or in defending the warren against aggressors. Vixen in *The Animals of Farthing Wood* likewise stays put in the earth while fox hunts (Dunn 2008, 176, 294). In *The Cold Moons*, the two females of the badger family even keep house for the patriarch of the family, prepare his bed (Clement 1988, 54) and provide his food, licking off the slime of snails to make them more palatable for him (87). Through their mildness and soft character these domestic females give psychological support to the males (e.g., Clement 1988, 106; Dann 2008, 244), and remain as a passive, but strengthening shadows at the backs of the brave heroes. Rebecca in *Duncton Wood* literally drains her life force into the hero Bracken to give him strength on his quest (Horwood 1980, 666), proving herself to be nothing but an extended part of him, perhaps his Jungian *anima*. Again, one can see at work the colonizing strategy of the master as identified by Plumwood (1993, 95): not only the animal, but also the female is absorbed by the masculine self. While the animal is assimilated as part of the self, the female is "commodified" (193) and all the parts that can "be made use of" by the male/master are ultimately "devour[ed]" (193). "Devouring," states Plumwood, "seeks to create a slave-world, a 'terra-formed' landscape which offers no resistance, which does not answer back because it no longer has voice and language of its own" (193). Nameless, voiceless, deprived of identity and agency, the female animal protagonist finds herself entirely dependent on the male without whom she drops out of existence. The stories act out patriarchal Utopian visions of "the implementation of the Cartesian dream of complete control over the other of nature" (193).

A framework of reallocated dualisms justifies the act of devouring the female. The male/animal is both natural and cultural, while the female/object in consequence can only be defined by negatives: she is un-natural and un-cultural. Indeed, left to their own devices, female characters in talking animal stories tend to do the wrong, the 'unnatural' thing. Dann's vixen only takes "a few seconds" (2008, 180) to come to the decision that she "want[s] to go wherever [fox] led" (180), yet these are sufficient "to put her life, and that of fox, in the gravest danger" (180), as her indecision exposes them to a fox hunt. The message is straightforward: if she would have submitted immediately, they would have escaped the danger. A similar construction can be seen in Tad Williams's *Tailchaser's Song* (1985). The cats who fall for the temptation of an 'unnatural' life under human care or in imitation of humans are the Queen of cats (152) and her court, as well as the hero's first love (360). That there is one female who does not fall for the temptation of the unnatural is self-evident: there needs to be a worthy mate to the hero. In *The Cold Moons*, the male badgers also have to protect a group of "weak-willed" (Clement 1988, 265) and "frail" (265) females, who, in best tradition of Eve, follow the lure of the unnatural Kronos.

"Reclaim[ing]" (266) the lost sows from Kronos is presented not as a superimposing of patriarchal power, but as an act of mercy, saving the ones that have "lost their way" (266). In plot constructions such as these, the "logic of the master" (Plumwood 1993, 190) comes full circle. In *The Cold Moons*, the unnatural female not yet devoured by the male/master almost shatters the badgers' dream of reaching their "Elysia" (Clement 1988, 56). In consequence, Clement presents it as an act of duty for the female others to pass agency to the male master as in their inherent unnaturalness they otherwise would become a threat to the community of animals as a whole—and thus, as the community is equated to an ecosystem, a threat to the environment. Other stories also present an unnatural female as a threat to the greater good of a self-constructed natural order appropriated to the hegemony of the master. This is a situation already pre-shadowed in *The Wind in the Willows*. As I stated above, there are no female animals in Grahame's world, but female humans do appear—and these, to the Riverbankers, "are even more threatening than the Wild Wooders" (Hunt 1994, 87).

To more thoroughly examine the nature of the "threat" (87) of the unabsorbed female in talking animal stories, it is necessary to take a step back and first view the political systems in the animal worlds. In consequence of the ultra-patriarchal society structures of most talking animal cultures, these are male-only domains, and additionally non-democratic ones. Democracy might appear as a temporary state, but the stable political solution is an elder council with a single male leader or a single male head animal in the position of an absolute ruler. Most stories take this for granted, employing the naturalization technique already sketched above: the political system is presented without any alternative or ideological struggle against it; it comes 'natural' to the animals.

As Blackford remarks, discussing Brian Jacques's *Redwall*, "peace [in the animal community] derives from stable gender roles" (2003, 102). In some stories, however, the stability of the system is threatened by females usurping power, an act of uprising of the other that is presented as so disturbing to the natural order that it even threatens the male/animal self with castration. *Tailchaser's Song* may again serve as an example. The cats are governed by a "Queen" (Williams 1985, 152) who has long given in to the lure of the unnatural. Her rule is opposed by the proud all-male "First-walkers" (88) who constantly reassert their masculinity through mock-fights and rough language (89) while they displace the fear of the consequences of the Queen's rule into their mythology. Their mythological hero—a deeply masculine extension of the self—meets a cat that defies classification in the either/or-structure of the male/female dualism. The cat turns out to be a castrated tom who is not only "half-fela," that is, half female (174), but in consequence also half-natural. Directly responsible are humans, not females, but as humans are the power behind the Queen's rule the story becomes an allegory for the consequences of female unnatural otherness gaining power over the masculine natural self. Defending masculinity thus is directly transformed into protecting the natural.

The connection between a threat to the natural and castration is even stronger in David Clement-Davies's *Fire Bringer* (2000). The story is of deer—animals whose life can easily be rendered as a patriarchal society. Yet the heroes on their journeys come across a herd ruled by vicious and ruthless hinds who force their stags into slavery while they assume all 'male' duties (335), including politics (336). In fact, even those females do not act on their own but follow the doctrines superimposed on them by another herd of males acting as *eminences grises*, but inside their own herd they reduce the stags to precisely the function normally given to female characters: they are put aside until "needed for mating" (335–336). The females have to continuously reinforce their power by 'unnatural' means: they force the males to remain in a location where the poisoned water of a copper mine cripples their antlers (335). The act of castration executed by destroying the phallus-symbol of the antlers merges with a threat to the environment originating from the same source—the poisoned water simultaneously pollutes the deer's habitat.

Naturally (another chimerical use of the term) it is the task of the hero and his friends to set right the 'natural balance.' This is achieved once the formerly crippled stag Haarg, now proudly bearing his restored phallus-symbol on his head (403), complacently nods to himself when observing the earlier lead hind as a compliant part of his harem (404), while all the traditionally male functions have been reassumed by the stags (403). The situation is almost exactly the mirror of the former one—only that the male rule, unlike the female one, needs not be reinforced by means threatening the environment. It is constructed as 'natural' and therefore right.

Peter Chippindale's *Mink!* (1995) follows a similar line. *Animal Farm*-like, the animals in his story are clearly metaphors for humans. The book understands itself as a witty satire on British politics of the 1990s. With the single exception of the rook Raka, female characters are either entirely domestic, plain stupid, or immensely annoying due to their feminist stance. Their influence in woodland politics is devastating. Fighting in absurd causes like "Worm's Lib" (85) females 'dabbling' in politics keep the animals (i.e., male animals) so busy that they are at first unable to cope with an invasive species overrunning their wood. When the invasive mink are finally defeated, the story makes it appear as if females virtually *have* to be stopped from having a voice in politics, as otherwise the wood would be exposed to new attacks threatening its balance. Boris Badger does away with the "threat" (Hunt 1994, 87) of the female with a hearty bite into the rabbit doe Cowslip's "prominent posterior" (Chippindale 1995, 543)—both a sexually loaded act and a demonstration of physical power. *Mink!* appears like a representation of the existential anxiety of androcentric Western ideologies exposed by the late 20th-century's challenge to patriarchal power structures.

The consequences, however, go beyond political systems. The restoration of the patriarchal hegemony also has immediate effects on ecological stability and the animals' connection to nature. Once the patriarchal system is restored, the threat of an ecological catastrophe is banned as well and the hero Owl suddenly experiences his wood as an epitome of peace and natural beauty and an eternal source of joy (564). In *Mink!*, but also in other stories such as *The Cold Moons* or *Fire Bringer*, patriarchy is thus directly linked to 'the natural' and, by means of employing the double-faced concept of nature, its normative goodness. Restoring patriarchy is equated with restoring the natural order, while restoring the natural order saves the environment from destabilizing factors. The stories present males as positively duty-bound to keep females in their place in order to keep the "river bank" in its Edwardian time bubble of eternal stability, both politically and, so the stories try to make believe, environmentally.

This connection between conservation and conservatism appears to be in a process of naturalization too. When Cosslett writes that there is environmental concern in *The Wind in the Willows* (2006, 180), it is precisely this naturalization which is at work. Neither Toad's motor cars nor any other technologies of the "Wide World" ever threaten the river bank environmentally. The threat is one to conservative culture, to patriarchal hegemony—and even Cosslett, in her otherwise admirable study, equates the two without questioning the connection.

Such presentations of females appear to stand in stark contrast to the women/nature analogy and the idea of the "angel in the ecosystem" (Plumwood 1993, 9) that Plumwood, writing in the 1990s and thus at the high time of talking animal stories, identified as "a popular contemporary green vision" (9). In this myth, the late 20th-century version of the 'angel in the house,' women appear as owning "special powers and capacities of nurturance, empathy and 'closeness to nature,' which are unsharable by men" (8). Several animal stories employ this concept and in doing so present a picture of the female animal that is vastly different from the dominant one sketched above, but equally serves to split the concept of nature into a male and a female domain. In William Horwood's *Duncton Wood* the character Rebecca appears not only as an angel, but also virtually as an avatar of the Ecosystem. Like in the image of women in *Country Women* that Padilla Carroll (2017) discusses, the "trope of the Goddess" is at work here. Rebecca is "personified Mother Earth" (Padilla Carroll 2017). Instinct-driven, naively innocent and able to even listen to the language of plants (William 1980, 137) the female mole is the connecting power between the male hero Bracken and the pantheistic energy underlying their world (664). Losing all subjectivity, she becomes a vessel for a universal Gaian love that needs to be transferred to the hero. Similarly, the she-wolf Larka in David Clement-Davies's *The Sight* is gifted with a form of super-instinct that allows her to touch pantheistic nature (2001, 396). Her brother is introduced as sharing her powers, but one never sees him using them in any comparable way. While she feels the pain of nature suffering under humans (475–476), he demands a pact with the humans to "protect the wilderness from [their] own power" (492). Again, she is the vessel, he acts. In both these cases nature appears as a double entity. There is a female nature, a Gaian spiritual power that is numinous and strange—the other—and a male nature, the material natural world as we experience it, an extension of the self. The latter is the one at which, in the ideology of the stories, environmental concerns should be directed, as the Gaian female sphere appears as unprotectable as it is unreachable.

When Baker claims that "something in the structure of the talking animal story makes it inherently subversive of patriarchal culture" (2001, 137), this unfortunately does not meet the facts for all talking animal stories. As, true to his introductory statement, "the animal can apparently be used to mean anything and everything" (4), it can also mean 'patriarchy.' Indeed, as the discussion above has shown, many talking animal stories transform the Arcadian idea of a unison of nature and culture into a patriarchal myth set in an idealized past. 'The animal,' as exclusively male as the concept of 'man,' continuously struggles to eradicate all challenges to the status quo, as the patriarchal system and an intact environment merge into one in a chimerical normative/materialist concept of 'nature.' The female is constructed as part of the problem, as an unnatural force invading from the outside, perhaps because emancipation of women and the environmental crisis began to stir patriarchal society around the same time and thus came to be associated by mere temporal coincidence. Through means of merging the normative and materialist dimensions of 'the natural,' some stories even make it appear as if the emancipation of women *caused* the environmental crisis, and, the other way round, that restoring Western patriarchal hegemony to its full strength would also restore the world's ecological balance. Other stories create visions of the 'angel in the ecosystem.' Again, females are not constructed as able to remedy the crisis—their connection is one to mystical, not to material nature. Agency again has to be passed to the male.

The genre with its high metaphorical potential and level of abstraction thus allows a platform to propagate values which are nowadays shunned in mainstream literature: patriarchy, in short, searches for a retreat on the river bank. Instead of attacking the foundations of the environmental crisis by a deconstruction of essentialist dualisms as Plumwood suggests in her groundbreaking *Feminism and the Mastery of Nature* (1993), the majority of talking animal stories operate inside Cartesian dualist thinking, seeking to dissociate nature from its connection with the female and instead associating it with the male, thus reinforcing the master narrative. Culture, too, remains firmly in male hands, or, in this genre, paws. Only the mystical, ungraspable sides of nature—those which material environmentalism cannot touch—are allowed to remain in the female sphere. The consequence is an even deeper trench between the dualist notions of male and female, whereby the latter category is stripped of almost all its contents. It may either remain an empty shell defined by negatives, whose sole purpose is reproduction, all its other meanings "devour[ed]" (Plumwood 1993, 192) by the totalizing self/master and "incorporated into the empire of the self" (192), or the female becomes an unreachable mystical 'other,' so alien that it does not have an identity but merges with a universal, but formless Gaian power. I follow Plumwood in her claim that such a restructuring of the content of Cartesian dualist boxes is not where any solutions to the environmental crisis will be found. Instead, Plumwood contends that one needs to go beyond dualism and recognize "both continuity and difference." In order to come to any understanding of the other, be it animal or female, the other needs to be "acknowledge[ed] . . . as neither alien to and discontinuous from self nor assimilated to or an extension of self" (6).

All these stories were written from the 1970s to the earliest years of the 21st century, in decades when feminist movements gained pace. It is certainly possible to argue that they reflect the male anxiety of the time, an anxiety which perhaps reaches its peak of expression in William Horwood's *Seekers at the WulfRock* (1997). In this Deep-Ecology-vision post-WWIII-Europeans in a dystopian year 2013 follow the examples set by the story's talking animals and return to a 'natural' life in unison with "our Earth, our Mother" (Horwood 1997, 7) which is highly praised by the narrator. These human characters live in "tribes" (7), ruled by "shaman[s]" (33), "elders" (37), and a "headman" (37) and women, silent, compliant, utterly passive creatures that are not even granted dialogue, are traded as objects among the men (39, 40).

As the genre virtually died with the beginning of the 21st century, it is difficult to discern a trend, but the tendency is at least hope-inspiring. In the isolated talking animal stories published during the last decade, one can see female characters invade plot structures as individual protagonists and gain cultural power as well as some leadership functions. But still the vein of utilizing 'the natural' remains in equal strength. While now more frequently including females, even contemporary talking animal stories simply elude other topics (females appear to be a very challenging one already). In all my readings, I have never come across a single talking animal story with a queer character. Talking animal stories apparently try to evade entering into a discussion of queerness, but in fact achieve the opposite: by ignoring queerness to the point of pretending it does not exist, the genre once again makes use of the blended physical/normative concept of nature. The stories present what they state to be (physically and normatively) 'natural' worlds. If in these worlds, queerness does not exist, the underlying implication is that it is unnatural—both physically and normatively. In this, talking animal stories reaffirm homophobic discourse even more drastically by evading the topic. The fight in the arena of 'the natural' has not stopped when finally granting more power to female characters—it simply shifted to other contestants.

References

Adams, Richard. 1974. *Watership Down*. London: Penguin.
Baker, Steve. 2001. *Picturing the Beast. Animals, Identity, and Representation*. Urbana, Chicago: University of Illinois Press.
Baran, Stephanie. 2017. "Visual Patriarchy: PETA Advertising and the Commodification of Sexualized Bodies." In *Women and Nature? Beyond Dualism in Gender, Body, and Environment*, ed. D. A. Vakoch and S. Mickey. New York and London: Routledge.
Barthes, Roland. 1974. *Mythologies*, trans. A. Lavers. London: Jonathan Cape.
Blackford, Holly V. 2003. "The Writing on the Wall of *Redwall*." In *Utopian and Dystopian Writing for Children and Young Adults*, ed. C. Hintz and E. Ostry. New York and London: Routledge.
Chippindale, Peter. 1995. *Mink!* London: Simon & Schuster.
Clement, Aeron. 1988. *The Cold Moons*. London: Penguin.
Clement-Davies, David. 2000. *Fire Bringer*. London: Macmillan Children's Books.
——. 2001. *The Sight*. London: Macmillan Children's Books.
Cosslett, Tess. 2006. *Talking Animals in British Children's Fiction, 1786–1914*. Aldershot: Ashgate.
Dann, Colin. 1979. *The Animals of Farthing Wood*. London: Egmont UK.
Derrida, Jacques. 2002. "The Animal that therefore I Am," trans. D. Wills. *Critical Enquiry* 28(2): 369–418.
Garrard, Greg. 2012. *Ecocriticism*. London: Routledge.
Grahame, Kenneth. 2008. *The Wind in the Willows*. London: Puffin.
Gruen, Lori. 2011. *Ethics and Animals: An Introduction*. Cambridge: Cambridge University Press.
Haraway, Donna. 1991. *Simians, Cyborgs, and Women: The Reinvention of Nature*. New York: Routledge.
Hawdon, Robin. 1984. *A Rustle in the Grass*. London: Hamlyn.
Horwood, William. 1980. *Duncton Wood*. London: Hamlyn.
——. 1997. *Seekers at the WulfRock*. London: Harper Collins.
Hunt, Peter. 1994. *The Wind in the Willows: A Fragmented Arcadia*. New York: Twayne.
Jacques, Brian. 2002. *Redwall*. London: Firebird Fantasy.
Manlove, Colin. 1999. *The Fantasy Literature of England*. Basingstoke: Macmillan.
Morton, Timothy. 2007. *Ecology without Nature*. Cambridge, MA: Harvard University Press.
O'Brian, Robert C. 1975. *Mrs. Frisby and the Rats of NIMH*. London: Puffin.
Padilla Carroll, Valerie. 2017. "Writing Women into Back-to-the-Land: Feminism, Appropriation, and Identity in the 1970s Feminist Magazine *Country Women*." In *Women and Nature? Beyond Dualism in Gender, Body, and Environment*, ed. D. A. Vakoch and S. Mickey. New York and London: Routledge.

Plumwood, Val. 1993. *Feminism and the Mastery of Nature*. New York and London: Routledge.

Moorhouse, Tom. 2013. "Praise for the River Singers." In *The River Singers*. Oxford: Oxford University Press.

Ratelle, Amy. 2015. *Animality and Children's Literature and Film*. Basingstoke: Palgrave Macmillan.

Stanton, Mary. 1988. *The Heavenly Horse from the Outermost West*. London: New English Library.

Sturgeon, Noël. 1997. *Ecofeminist Natures: Race, Gender, Feminist Theory and Political Action*. New York: Routledge.

Teodorescu, Adriana. 2017. "The Women-Nature Connection as a Key Element in the Social Construction of Western Contemporary Motherhood." In *Women and Nature? Beyond Dualism in Gender, Body, and Environment*, ed. D. A. Vakoch and S. Mickey. New York and London: Routledge.

Valera, Luca. 2017. "Françoise d'Eaubonne and Ecofeminism: Rediscovering the Link between Women and Nature." In *Women and Nature? Beyond Dualism in Gender, Body, and Environment*, ed. D. A. Vakoch and S. Mickey. New York and London: Routledge.

Williams, Tad. 1985. *Tailchaser's Song*. New York: Daw.

3 Visual patriarchy: PETA advertising and the commodification of sexualized bodies

Stephanie Baran

Advertising is a multifaceted idea that is utilized by every single group, company, organization and institution to get their product out to the public. The goal is to get the public thinking about the message or product and hopefully buy-in and support. Capitalism is the driver of the advertising world, because it does not matter if the organization is for-profit or not, money is the major component. And most advertisers rely on the old adage, 'sex sells.' For example, any perfume/makeup/clothing advertisement is selling an emotion, feeling, or self-presentation or image to the viewer. However, if we shifted our thinking to social justice organizations, and more specifically environmental justice organizations, one would think that these same 'sex sells' mantras would not exist.

Advertising has been utilizing the frame of 'sex sells' for some time. This often incites a knee-jerk reaction, or gets the viewer to stop and examine their work for a few seconds. In relation to animal rights advertising, researchers find that the particular imagery of 'sex sells' has a purpose: to get people thinking about animal cruelty by putting women in the place of animals to highlight the mutual connection (Lunceford 2012; Olson and Goodnight 1994). However, this analogy fails because in patriarchal culture women are meat and are to be consumed in a variety of ways. Therefore, advertising is, in a sense, visualized patriarchy—the actual visualization of patriarchal ideas and social norms. Therefore, Adams (2010) finds that positing women in the place of animals makes violence the 'absent referent,' which acts as the disconnect between the beginning and end. In sum, in both situations, women and animals experience a similar violence within humanity. While PETA wants to show the horrifying treatment that animals receive, using women to do it is a failed analogy because under patriarchy, women are meat.

However, the People for the Ethical Treatment of Animals act in a paradoxical nature when considering this very issue. This organization prides itself on protecting animals and showing the trauma and violence that animals face, and yet uses women and men to accomplish those goals. However, when considering the difference of presentation of the ads that feature men and women, there are cultural scripts infused into these ads. For example, PETA advertising desires "rebelliousness" from meat-eating but also values conformity to cultural desires—like being a certain body type, look, etc. (Bordo 1997). Often when we think about patriarchy, it is in a manner, such as the gender pay gap (solely represented as between men and women and never considering within-gender gaps) or the way that men and women are differently treated by Instagram, etc., and often the visualized aspects of patriarchy in advertising are not examined. Many of these ads feature men and women pictured differently. For example, men are usually photographed in positions of power and, conversely, women are photographed in positions of sexualized vulnerability. The visual patriarchy used by PETA within their images does not just use gendered scripts about men and women, but also transphobia, fatphobia, racialized scripts, and ableism. Visual patriarchy most directly fixates its position on how women, and in many cases men, feel about their bodies and their own body image (Mitten and D'Amore 2017). Therefore, this works to detail how patriarchy is multifaceted. And by trying to elevate the position of animals, it denigrates the positions of routinely victimized populations.

Advertising uses women in the same way in which the food and clothing industries use animals: as commodities. Advertising uses the capitalist frame to posit women as consumable objects for the consumption of others—in this case, the cismale, heteronormative gaze. Therefore, the purpose of this chapter is to illuminate the ways that all advertising utilizes these frames and gazes, and as scholars, activists, and humans existing under a variety of conditions, we should be aware regarding the ways in which intersections of oppression function under the capitalist tendencies that are exhibited through advertising.

PETA uses imagery and cause to incite a reaction, but not in a revolutionary way or as a counter to the established frame. PETA's states their mission as "animals are not ours to eat, wear, experiment on, use for entertainment, or abuse in any other way" (PETA.org n.d.). While this statement is their mission statement, in order for it to get to the public, they must manufacture the consumption of sex to sell their products and also motivate people to support their mission. Furthermore, to communicate with the general populace, PETA must use the language of the dominant group: the established patriarchal male gaze (Julius 2012). More importantly, in order for PETA to be present in the collective minds of the general populace, they must 'sell' women in order for people to even give a thought about animals (Deckha 2008).

Does the messaging and visual patriarchy used by PETA detract from their purpose and harm women in the process? The paradoxical nature of a seemingly leftist organization such as PETA would assume that they would understand this positionality; however, this is an incorrect explanation. To dive deeper, the basis of this exploration stems from a content analysis of 100 photos from the PETA.org website from five different media campaigns, including ableist, trans and fat phobic PETA ads and videos. The videos are an avenue for animal rights outreach by celebrity activists, and the videos use women to *tantalize* and lack a serious discussion, information or any visual representation of animal torture, etc. In contrast, videos featuring individuals who identify as male celebrities involve a clothed man *informing* the public, with violent images and facts. In these videos, all femininity is removed and the videos are explicitly serious and masculine. Therefore, PETA posits men as having facts and information and women being available for sensual titillation, visual pleasure and no real substance to why caring about animals and being vegetarian/vegan is important.

Women as consumable objects

Debord (1994) illustrates that the spectacle in this case advertising, is the "*outcome* and the *goal* of the dominant mode of production [emphasis added]" (13). More in-depth, Debord continues by stating that no matter the mode—news sources, marketing or the entertainment industry, "the spectacle is the [epitome of] the prevailing model of social life." To understand advertising and the use of women in those particular images, it is important to remember that all oppressions are interconnected, which at its base is the foundation for conflict theory. For example, Patricia Hill Collins (1993) notes that while existing in this particular society, we are at once oppressors and oppressed.

The intersectionality of race, class, sexuality, and gender oppression all converge within the advertising world. Collins explains that because we do not live in a vacuum, the intersectionality of race, class, sexuality, and gender construct the different ways in which we view the world. For example, Collins explains "how sexualities constructed in conjunction with an unquestioned heterosexism become manipulated within class, race, and nation as distinctive systems of oppression" (2000, 132).

Advertising campaigns of all kinds construct class and racializing distinctions within their outreach. The essence of advertising is to give the public the need to have the product that they are selling. For example, perfume ads do not have anti-social people ignoring each other because they do not want to talk to them. These ads feature a high-class statuses and lifestyles that people can desire. These products create a fetishism that connects consumers to the want and potentially the connection to that lifestyle. This fetishism then extends to the people that appear in the ad. Therefore, not only are products fetishized but also the people in the ads are further fetishized and essentialized.

Essentialism assumes that in relation to any concept, there are a set of characteristics that must be possessed and whatever that is must belong to the essence of that concept (hooks 1991). Therefore, the essential qualities of women include the placement of their bodies and all women must have those essentializing qualities, and if they do not, then their bodies are somehow different. Advertising therefore creates essentialized bodies for consumption by society within patriarchy and therefore extends these characteristics to every woman. For example, advertising often features parts of women and therefore makes them consumable parts—an arm with a perfume bottle, or a leg with a pair of shoes. As Julius states, women are more likely to be featured 'in pieces' and a woman becomes 'parts' versus a whole woman (Julius 2012; Ward, Merriwether, and Caruthers 2006). The male gaze presented in patriarchal society only sees women as pieces and not whole people; therefore, it is far easier to treat women as second-class citizens only existing to please heterosexual, cismen.

Other research has found that this dual representation, both woman and pieces of women, reasserts speciesist ideology and reinforces both anti-woman and anti-animal sentiments (Glasser 2011). This visualized patriarchy is the mantra of sex sells within the capitalist system. As hooks (1997, 8) notes, "issue is not freeing ourselves from representation. It's really about being enlightened witnesses when we watch representations, which means we are able to be critically vigilant about both what is being told to us and how we respond to what is being told." Representations hold a social value and often maintain that value. For example, sensual-looking women posing for perfume advertisements or a passively dominant man posing for Calvin Klein advertisements all blend to culturally manufactured images (Bordo 1997). Both hooks and Bordo build on Lorde's power of the erotic to explain the influence that the manufactured body holds within media (Bordo 1997; hooks 1991, 1997; Lorde 1984).

Body policing as consumption

For advertising to remain profitable, a continual source of consumable objects is necessary. Therefore, body policing becomes as suitable avenue for a highly profitable revenue stream, which includes shaving, facial, skin and weight loss products. Continuing with the social justice paradox and how we think of 'left leaning' organizations, we might assume that PETA would be above body policing ads. Collectively, throughout all the ads and videos under the PETA brand (because ostensibly, that is what PETA is), there was a connection to body policing in relation to grooming, shaving, and general 'attractive appearance.' For example, ciswomen and LGBTQAI men were depicted as unattractive if they did not shave their bikini area or their underarms. These ads advocate for no fur trim on coats and other apparel in order to protect animals that usually see violence—minks, foxes, and other small furry animals.

In connection to policing body hair, a few of the ads feature women who are not perceived as the 'skinny standard.' For example, in an ad that featured a woman outside the normalized 'skinny' body type within current society, her body was placed in such a way that her placement hid her stomach and Photoshop was used to disguise her body with highlights and shading. This body positioning is different from the featured models in the other campaigns where their bodies and silhouettes are clearly visible. While there was an attempt to use a model that was different than a majority of their other advertising campaigns, there was still a conscious attempt to hide that body. Therefore, the paradox of a left-leaning organization to be somewhat body positive, but is perhaps not, signifies another concern within this particular social justice organization.

Continuing with problematic body policing, a number of ads that feature meat eating compare fatness to physically embodying a pig. Therefore, this creates the illusion that all vegans and vegetarians are skinny and sexy and that is the only acceptable body, thereby erasing all other body types. This erasure couples fatness with meat eating, excess and shame and as something that should be shaded and hidden (Bordo 2004). These ads promote ablest thinking by equating fatness to lesser intelligence and outright stupidity. Therefore, by trying to get their message out that we as humans should not be using animals for our benefit, they actually oppress humans in the process.

The desire to reaffirm what people think is 'sexy' or attractive or *how* to appear as such is explicit within advertising and sadly is no different in regards to PETA's advertising. The ads for men focus on sexual prowess and that being vegan would result in an easier time finding women. The lifestyle of vegan/vegetarianism is viewed as 'sexy' in PETA's eyes and therefore women will find you sexually appealing if you identify with this lifestyle. In opposition to that, ads featuring women use the premise of sexual availability under the guise of veganism. It's about bodies and *being* a certain type of sexy. The focus is on sexual interest and desire and the side benefit is that you might help an animal or two if you went meat-free.

The body policing continues with ads that focus on facial acne, specifically using ads featuring women. Therefore, the attention to bodies and how people present themselves to the world and the ways in which PETA thinks vegetarians and vegans *should* look is problematic. Instead of the focus being on how one can best help animals, the focus is the continued preoccupation with being the 'best' self—which culminates in a good-looking self as a good-doing self (Brumberg 2010). Thus, the focus on the body continues the problematic struggle that women and men face with body issues and the social controls on those bodies. PETA continually focuses on what society deems as presentable and does not challenge the established social norms about how people should present their bodies. Understanding intersectionality requires that PETA and society realize that human rights are animal rights and vice versa.

Sexualized shock and awe advertising and the male gaze

Capitalism effectively sells the bodies of women for consumption for the masses. Debord notes that "[b]y the means of the spectacle the ruling order discourses endlessly upon itself in an uninterrupted monologue of self-praise. The spectacle is the self-portrait of power in the age of power's totalitarian rule over the conditions of existence" (1994, 19). Therefore, this is an expansion that the ideas of the ruling class are the ruling ideas of the age (Marx and Engels 1965). Therefore, the ideas regarding advertising and the consumption of bodies in media are representative of how the ruling class views these particular participants in society.

In regards to creating a media frenzy, the idea of the public spectacle is important in order to solidify an attention-grabbing technique (Donovan 1990; Julius 2012; Lunceford 2012; Matusitz and Forrester 2013; Olson and Goodnight 1994; Simonson 2001). Lunceford (2012) examines the variety of media tools that PETA uses to engage the public. For example, he focuses mostly on PETA's "Running of the Nudes" event every year in Spain. This correlates with the Pamplona running of the bulls. Lunceford claims that the collection of naked bodies is the moment where "humans cast off their humanity and stand as one with animals" (2012, 8). The problem with this assumption is at no point in the current era are bodies not seen in a sexualized manner (except of course, during and after pregnancy).

Whenever women's bodies are nude, the representation of their body is *protected*. In contrast, topless men never are depicted with blue ribbons across their chests. Therefore, this compounds the idea that the female body is something special and treated differently. For example, the campaign "Free the Nipple" seeks to end the gendered oppression and inequalities that women experience given the overt patriarchal policing of women's bodies which differs from male bodies in present society. In more academic research, Ward et al. (2006, 704) finds that "dominant gender ideologies, as exemplified by the media, consistently paint women as sexual objects, highlighting their bodies as being mainly for the pleasure of men instead of as multidimensional (i.e., including both reproductive and sexual functions)." Further, this research found that exposure to the media's sexualized images of women led men to hold more sexist and gendered views about women and their bodies.

Sexualized imagery

Advertising relies on sexualized imagery and extends visual patriarchy, and based on the extensive research that has emerged about PETA, we would be remiss if we thought PETA was any different. There is no problem with women being sexy, but there is a problem if men do not equally share that same sexualized nature, which will be unpacked shortly. Many of the images relied on overt sexual innuendo or mostly naked women to sell the slogans of PETA. These images use women to associate *male* sexual performance with meat eating. For example, if you eat meat, you have a higher likelihood of being impotent by some unconfirmed research. These images also explicitly detailed that meat eating would have men more likely to experience less energy during sex and have less energy during the experience. The key to all of these images is that they use women and put the focus on heteronormative, cismale sexuality. The women in these ads are only present for male enjoyment.

Similar to the advertisements, videos used by PETA depict women as a solely sexualized being as the foil for male enjoyment and male sexuality. The video "Veggie Love" depicts three models in lingerie rubbing a variety of vegetables on their bodies. These white models depict one particular body type—skinny. It appears that the PETA objective was to pose these particular models in a similar way to a Victoria's Secret advertisement. The main focus of the video is script that says, "Studies show that vegetarians have better sex." Here, the idea is not to become a vegetarian for the animals' sake, but to become a vegetarian so that there might be better sex.

In connecting visual patriarchy with the spectacle, PETA takes the crude idea of jelly wrestling, changing it to tofu which makes it far more animal friendly.

The reason that this connects to animal rights is due to gelatin being an animal by-product. The women in this advertisement are wearing bikinis and wrestling in tofu to bring awareness to animals! The advertisement notes how many people are stopping and gawking but the idea of the spectacle is really what is drawing the focus. More deeply, people stopping are not particularly interested in the fact that these women are dedicated to the cause of animal rights, but that they are wrestling in bikinis in the middle of a busy intersection. All these spectators are viewing is a spectacle and not the meaning behind the spectacle.

Sexualized violence

PETA uses images of women being beaten, abused, and disturbingly sexualized to elicit a response that would connect this violence to the violence against animals (i.e., look at this beaten woman, how could you let this happen to an animal?). However, this is a faulty analysis because this violence happens to women on a daily basis. For example, in real life women are beaten, stabbed, or murdered for rejecting male advances. Therefore, PETA wants us to view the image of the woman, but in the context of animals, but never challenges that women also experience similar violence on a daily basis. For example, one ad showcases a woman in a fancy dress out to a nice dinner and a man with a glove is choking her while imitating how the process of foie gras is completed. There are also images of women as the varieties of meat cuts, displayed as hanging meat, in meat trays, etc. Here, women are no longer whole people, but just pieces of flesh that are removed from their humanity.

Women can experience violence on a daily basis: from walking away from an annoying catcaller to screaming for her life from someone attacking her because she said no. Women can recount a variety of these experiences over their lifetime. Instead of highlighting the violence that women experience, PETA decides to take a different approach to sexual violence. This video begins with a woman in just a bra, pair of panties and a coat, walking down the street, looking sad and walking as if in pain and in a neck brace. The woman walks into the apartment and the boyfriend is fixing a hole in the wall. The voice-over indicates that her boyfriend went vegan and knocked the bottom out of her and this was okay because she liked it—which is apropos in rape discourse. Therefore, in this video, PETA highlights that paradoxical nature—you do not have to be conscious of violence within humanity to be a vegan. The idea of consent is missing here as well. For example, the woman may have consented to sex in this video, but perhaps not the style. PETA uses rape discourse to highlight that you can have rough sex as a vegan. This video continues to highlight the heteronormative *men's* sexuality at the loss of woman's. It displays still men as predatory and women as harmless bystanders available for male consumption. The video never looks at the woman as sexually empowered only as an innocuous victim. The video frames sexual abuse as normal and therefore a glorification of domestic violence.

Violent images

PETA relies on shock value—which seems so easy and often is questionable to their cause. For example, these images of shock value include women and *some* men in situations that exhibit death, are violent, and include murderous acts. Images include women being dismembered and hanging on meat hooks, on a coroner's table, complete with a toe tag, and the tagline reads, "I'd rather be dead, than wear fur." Other images categorized as violent are images where naked women appear to be dismembered and disfigured in Styrofoam containers, like how meat is sold. Other images include naked women, painted red, in an 'oven' with the tagline, 'McDonald's Burns Chicks Alive,' as a play on words for 'Chicks,' that is, women to stand in for baby chickens. Here there is a multiplicity of things being demonstrated. Not just the idea that these women are naked to 'make a point' but that women are referred to as 'chicks' which has the connotation of being innocent and infantile. Therefore, women have the dual role of innocent baby animal and at the same time sexualized woman. While all of these images are horrifying and are related to ways that animals are handled in order to be consumed by the general public. It also is related to how women are viewed as objects to be preyed upon.

Racialized systems

PETA often intersects gender and race, often with a racialized discourse. For example, for a PETA ad condemning wild animals, the ad features a muscular, naked black man, with his mouth open in a roar, trapped in a cage in a very animalistic body positioning. PETA uses several racial tropes, first with the black and brown men as animals—therefore not human. This should really give us pause given the recent, but not new revelations regarding the treatment of black men and women. What is important about this is that white men do not appear in the same way as the men of color. One ad depicts a boa wrapping themselves around someone, but white men by and large are not pictured in the same manner. Given the way that black men are demonized in society by white supremacy, this image should initiate an immediate response.

Another ad featuring a woman of color included the tagline, "Dogs need birth control too!" For example, as Roberts argues in *Killing the Black Body: Race, Reproduction and the Meaning of Liberty* (1997), that black and brown women have had their reproductive choices limited for them and in many cases chosen for them. However, this image is completely problematic given the history of eugenics and racist reproductive rights policies for women of color—in relation to Clinton's welfare 'reform' women of color, particularly black women, in legislation that tried but ultimately failed to have the Mirena and Norplant (IUD) implanted to receive welfare benefits (Collins 2000; Roberts 1997). Images like this thereby treat women of color as second-class citizens *and* are rather insensitive to women who have had to endure reproductive violence. For example, in one such media campaign, a model of color was placed as an "Eve-esque temptress," which correlates to the idea of black women as jezebels or Latinas as 'hot-blooded' or overly sexual (Collins 2000, 2004; Roberts 1997). It also reduces women to stereotypes that have been used by white supremacy to control women of color—jezebel, baby producers, overly fertile, sluts, etc. (Collins 2000, 2004; Roberts 1997). Because PETA is using the language of the dominant (white, heterosexual cismales) group, the images and videos produced by PETA use subordinate members of society in a particularly marginalizing way.

Pieces of 'woman' and the heterosexualization of the body

Lee (1994, 343) explains that "the body is a 'text' of culture; it is a symbolic form upon which the norms and practices of society are inscribed" (cf. Fraser and Bartky 1992; Haug 1999; Jaggar and Bordo 1989). As stated previous, the positions that bodies encompass within advertising illustrates how the dominant frame understands those bodies within society. This dominant norm gives meaning to these particular bodies, and if they are in pieces, then the people attached to those bodies lose their meaning. PETA creates a type of 'state of exception' on women's bodies when presenting them to men (Agamben 1998). They accomplish this by making the visualization of violence of animals on women and thereby acting as a sense of state—animals should not by hunted for game, but we will display 'women as game' by using women in violent positions. For example, PETA opposes hunting but actively uses women for sport (in advertisements, tofu wrestling, pillow fights, etc.) to garner visual spectacles and interest from heterosexual men.

PETA tried two different marketing techniques. One of their marketing campaigns used visual images of animal torture, but of course, no one wants to confront or be confronted with the threat of humanity upon animals (Julius 2012). The opposite was to use two women in bras and panties engaging in a pillow fight to get the attention of men to "Go Vegetarian" (Matthews 2007; Julius 2012). Matthews, the senior vice president of PETA, finds that "women are often the only ones with the 'cojones' to put themselves on the line for their beliefs" (Matthews 2007, 215). This statement is using a euphemism for male genitalia which is rather interesting when discussing the 'empowerment of women' from a male perspective (Julius 2012).

Audre Lorde's power of the erotic comes to mind in reference to Matthews's quote. Ad campaigns are facilitated from an internalized male gaze, not from the power of the erotic. He also lacks the ability to recognize where the idea of women joining these nude photo shoots are in relation to their subconscious or conscious pressure to fit within the ideals of the male gaze, hence suppressing the erotic for the substitution of the pornographic. Lee (1994) notes that the male desires of woman and about women have been encrypted onto their bodies. Lorde states that "the erotic has often been misnamed by men and used against women. It has been made into the confused, the trivial, the psychotic, the plasticized sensation" (1984, 54). So the women in the photos may be loving what they are doing with intense passion, but that is what Lorde is arguing is the erotic. For example, Alicia Silverstone exhibits Lorde's notion of the power of the erotic best in her PETA video. Silverstone is a well-known vegan and has posed for PETA in previous ads. Here, she is seen swimming naked in a pool advocating for animal rights, because somehow those things make sense. The main connection here to animal rights and bodies is that 'she feels better' and the only way to get this across to the public is to emerge out of a pool, naked, with a slow beat in the background.

Patriarchy and men

This analysis used images that featured men as well. The interesting aspect of these images is that all the men had trusting expressions. Therefore, there is a disconnection that men are ever part of a system that might hurt women. Another aspect was that most of the images that featured men pictured them as fully clothed and with live animals. In the ad campaigns analyzed, women were almost never pictured with live animals. Also, the taglines were also very benign, which differed completely from the taglines that featured women. For example, these ads had statements like, "I'm Tom Higgenson and I'm a vegetarian." The ads feature men within a heteronormative lens which is focused on attracting female attention, with taglines such as, "Chicks dig men who are vegetarians." The comparison of women and animals with the word 'chick' should give us pause.

PETA effectively reaffirms socially constructed and gendered norms about food consumption, particularly regarding meat and vegetables. Their positing for 'strong men eat veggies' technically reaffirms the masculine identity associated with meat (Adams 2010; Luke 2007). Connected to that are images of women becoming vegetables or being hung like meat—both a consumable object, subconsciously reminding the viewing public that these images, food, *and women* are to be consumed by men. Their argument is framed in a way that when taken together, creates a hostile comparison for women.

While some of the ads did include naked men, the position of the body is extremely important. The torso is the most vulnerable part of the body for humans and animals alike. Therefore, when animals show you their torso, it shows trust or submission. In the ads that included naked men, they differed from the ads with women in a considerable way. In ads that include naked women crouching, their skin was branded, scratched, and bleeding with the tagline "beaten, shackled and abused." In similarly positioned men's ads, the tagline was "no animal belongs on chains" and the male model was not featured with any scratches or open wounds, and was standing up, demonstrating strength and resilience. These ads were very focused on strength and a particular type of hegemonic masculinity. There are many types of masculinities and these ads focus on strength and the ability to attract women in a specifically cisheteronormative way and disregards the multitudes of masculinities that can exist (Connell and Connell 2005; Connell and Messerschmidt 2005; Hennen 2008).

Meat is socially constructed as manly, therefore PETA runs ads that focus on vegans who are strong and work body builders to assure men that if they stop eating meat they would not be seen as weak (Adams 2010; Luke 2007). Therefore, the focus is still on the identity of cis, heterosexual hegemonic masculine strength. Another key piece found in the analysis is that men of color overwhelmingly make up the proportion of naked and exotified advertisements for PETA. However, if ads do include naked men, they are never depicted in a sexual nature. For example, in ads for "Ink not mink," sports stars have strategically placed footballs, but are not positioned in a sexual way, but in a position of strength. In order to get this across, vegan Casey Affleck talks seriously about the issues around animal torture—specifically the dehorning of animals. Here, Affleck discusses the severity and utter cruelty that animals endure at the hands of human beings. He talks seriously to the camera, fully clothed as violent images detailing this process scroll behind him. The tone is serious, filled with information and facts about the dire situations in which humans place these animals. Therefore, in contrast to the Alicia Silverstone video and aligning with all of the other images and videos, again, men have information and are serious.

The interesting thing is that all of images and videos of men are also complicit in patriarchy. While working differently on men, specifically white men benefit from patriarchy. White men are photographed as smiling, looking fresh, and never enacting violence. White men, in particular, are not used in a manner that does nothing if not reaffirm the hegemonic white heteropatriarchy. Black men, however, do not benefit from patriarchy in the same manner. Black and brown men are depicted as super strength, reacting or in animalistic position. The reason why this is related to patriarchy is that they do not benefit from patriarchy in the same way that white men do.

Intersectional environmental justice: beyond depictions of ableism and transphobia

One of the more surprising finds of this project are the ads that include ablest and transphobic messages. These messages include a false connection between dairy consumption and autism and transphobic messages with men and body types—specifically included the tagline "Dude looks like a lady," not only negatively referencing a man who is fat and also condemning transgender people in the process. Again, within these advertisements there is a dedication to normalized heterosexuality. There are no other deviations from gender, solely men and women and people are taut, skinny, and not fat.

Other ads by PETA advocate that meat eating reduces intelligence; one might become 'a vegetable.' Unsurprisingly, PETA also has advertisements where *women become* vegetables. These ads create a thought dissonance—you can be 'manly' by eating vegetables and you can have better sex (but your partner might suffer) and eating meat is ruining your intelligence. When viewed together, these ads are making a statement that reaffirms 'meat as manly,' but don't lose your masculinity by becoming a vegetable if you eat meat. Conversely,

PETA has ads where women *already are/are becoming* vegetables. In this instance, women can only *be feminine*. If men become vegetables, they lose all their essential qualities as men, and therefore become effeminate, which for PETA and generally accepted social norms—is bad.

In order to build a better organization, we must bridge the gap between more radical organizations and PETA. As social justice activists, scholars, and people existing under a variety of conditions, we should always remember that our commitment is not just to human beings, but also to all beings and it must be truly intersectional. Being congruent and conscious is not easy and there are all sorts of aggressions that take place in our lives that make thinking about social justice and our place in the world difficult. Gorski (2014) notes that macro-aggressions are the ways in which larger structures impact lives in different ways and construct our decisions. And as always, as activists, we must remember that we as people do not operate in a vacuum and that all of our actions must be as conscious and congruent as possible.

PETA actively excludes individual people from engaging within their cause due to its problematic messaging. If you fail to fit into PETA's image of skinny, able-bodied, heterosexual and white, you are their target. PETA's messaging erases the visibility of marginalized communities and potentially makes them feel unwelcome in the vegetarian and vegan community.

While not all of PETA's ads are misogynistic and violent, the fact that even one of these ads exist should give us pause. Therefore, while PETA claims to be ridding the world of oppression against animals, it uses women to frame that oppression. The problem with this analogy is that patriarchy does kill and does treat women as pieces to be exotified. PETA does not reclaim 'sexy,' counter hegemony, or anything of that nature. PETA plays to these issues that women are to be overtly sexualized and objectified because in our culture that will give them some press. Given the oppression and domination over women and their image, displaying women in shackles, beaten, dismembered, and displayed in meat packing trays actually hurts women because these things actually happen. Therefore, the message that PETA tries to send is lost because patriarchy actually creates this system of oppression upon women in the same way that capitalism and factory farming create a system of oppression on animals. The advertising of PETA therefore reaffirms patriarchy by presenting it in full view and by society not being able to connect that what happens to animals, happens to women and to some men.

PETA and other organizations like it must fully realize their position to challenge patriarchy and fully engage with environmental justice. They must understand the positionality of women and men in their advertisements and be fully committed to intersectional environmental justice.

References

Adams, Carol J. 2010. *The Sexual Politics of Meat: A Feminist-Vegetarian Critical Theory*. London: Bloomsbury.

Agamben, Giorgio. 1998. *Homo Sacer: Sovereign Power and Bare Life,* trans. D. Heller-Roazen. Stanford: Stanford University Press.

Bordo, Susan. 1997. *Twilight Zones: The Hidden Life of Cultural Images from Plato to OJ*. Berkeley: University of California Press.

——. 2004. *Unbearable Weight: Feminism, Western Culture, and the Body*. Berkeley: University of California Press.

Brumberg, Joan Jacobs. 2010. *The Body Project: An Intimate History of American Girls*. New York: Random House.

——. 2004. *Unbearable Weight: Feminism, Western Culture, and the Body*. Berkeley: University of California Press.

Collins, Patricia Hill. 1993. "Toward a New Vision: Race, Class, and Gender as Categories of Analysis and Connection," *Race, Sex & Class* 1(1): 25–45.

——. 2000. *Black Feminist Thought: Knowledge, Consciousness, and the Politics of Empowerment*. New York: Routledge.

——. 2004. *Black Sexual Politics: African Americans, Gender, and the New Racism*. New York: Routledge.

Connell, Robert William and Raewyn Connell. 2005. *Masculinities*. Berkeley: University of California Press.

Connell, Robert W. and James W. Messerschmidt. 2005. "Hegemonic Masculinity Rethinking the Concept," *Gender & Society* 19(6): 829–859.

Debord, Guy. 1994. *The Society of the Spectacle,* trans. D. Micholson-Smith. New York: Black and Red Press.

Deckha, Maneesha. 2008. "Disturbing Images: PETA and the Feminist Ethics of Animal Advocacy," *Ethics & the Environment* 13(2): 35–76.

Donovan, Josephine. 1990. "Animal Rights and Feminist Theory," *Signs* 15(2): 350–375.

Fraser, Nancy and Sandra Lee Bartky. 1992. *Revaluing French Feminism: Critical Essays on Difference, Agency, and Culture*. Bloomington: Indiana University Press.

Glasser, Carol L. 2011. "Tied Oppressions: An Analysis of How Sexist Imagery Reinforces Speciesist Sentiment," *The Brock Review* 12(1): 51–68.

Gorski, Paul C. 2014. "Consumerism as Racial and Economic Injustice: The Macroaggressions That Make Me, and Maybe You, a Hypocrite," *Understanding and Dismantling Privilege* 4(1): 1–21.

Haug, Frigga. 1999. *Female Sexualization: A Collective Work of Memory*. New York: Verso.

Hennen, Peter. 2008. *Faeries, Bears, and Leathermen: Men in Community Queering the Masculine*. Chicago: University of Chicago Press.

hooks, bell. 1991. "Essentialism and Experience," *American Literary History* 3(1): 172–183.

hooks, bell and Sut Jhally (director). 1997. *bell hooks: Cultural Criticism and Transformation*. Transcript of the film. Northhampton, MA: Media Education Foundation.

Jaggar, Alison M. and Susan Bordo. 1989. *Gender/Body/Knowledge: Feminist Reconstructions of Being and Knowing*. New Brunswick: Rutgers University Press.

Julius, Marina Elizabeth. 2012. "Nonhuman Animal Appeal: An Ecofeminist Exploration of PETA'S Business Model." MA thesis, San Diego State University.

Lee, Janet. 1994. "Menarche and the (hetero) Sexualization of the Female Body," *Gender & Society* 8(3): 343–362.

Lorde, Audre. 1984. *Sister Outsider: Essays and Speeches.* Berkeley: Crossing Press.

Luke, Brian. 2007. *Brutal: Manhood and the Exploitation of Animals.* Urbana: University of Illinois Press.

Lunceford, Brett. 2012. *Naked Politics: Nudity, Political Action, and the Rhetoric of the Body.* Lanham: Lexington Books.

Mathews, Dan. 2007. *Committed.* New York: Atria Books.

Marx, Karl and Frederick Engels. 1965. *The German Ideology.* London: Lawrence & Wishart.

Matusitz, Jonathan and Maya Forrester. 2013. "PETA Making Social Noise: A Perspective on Shock Advertising," *Portuguese Journal of Social Science* 12(1): 85–100.

Mitten, Denise and Chiara D'Amore. 2017. "The Relationship of Women's Body Image and Experience in Nature." In *Women and Nature? Beyond Dualism in Gender, Body, and Environment,* ed. D. A. Vakoch and S. Mickey. New York and London: Routledge.

Olson, Kathryn M. and G. Thomas Goodnight. 1994. "Entanglements of Consumption, Cruelty, Privacy, and Fashion: The Social Controversy over Fur," *Quarterly Journal of Speech* 80(3): 249–276.

PETA.org (n.d.) www.peta.org/features/12-petas-creepiest-ads/#ixzz3KEk5y5dW

Roberts, Dorothy E. 1997. *Killing the Black Body: Race, Reproduction, and the Meaning of Liberty.* New York: Pantheon.

Simonson, Peter. 2001. "Social Noise and Segmented Rhythms: News, Entertainment, and Celebrity in the Crusade for Animal Rights," *The Communication Review* 4(3): 399–420.

Ward, L. Monique, Ann Merriwether, and Allison Caruthers. 2006. "Breasts are for Men: Media, Masculinity Ideologies, and Men's Beliefs about Women's Bodies," *Sex Roles* 55(9–10): 703–714.

4 Ethical transfeminism: transgender individuals' narratives as contributions to ethics of vegetarian ecofeminisms

Anja Koletnik

This chapter will stress feminisms as being primarily commitments and projects of ethical and political inclusivity and solidarity, rather than forefronting them as projects of advocacy for binary gender categories and gender identity politics. Exemplification of my argument will be built upon Carol J. Adams's canon of vegetarian ecofeminisms, who appropriate diet as their essential resource of political work with which they contest speciesism, cissexism, and patriarchy.[1]

'The personal is political' is a resonation present more or less distinctively within all feminist discourses, vegetarian ecofeminisms, and their conceptualizations of food being no exception. 'Food viscerally connects individuals and social bodies,' claims Belasco, and suggests the 'politics of food, culture, and society is an increasingly important area for social inquiry' (Belasco 2008 in Cherry 2006, 233). Cherry (2006) implies that consumption identities transcend the initial understanding of the concept of identity as being what someone is, and build upon what someone does. Discursive realms are thus being transcended with considerations of identifications based on the materiality of lived experiences—a central stance of my argumentation, which will correlate diet and gender as analytical categories manifested upon the example of transgender and cisgender non-conforming individuals' meat non-consumption.[2]

Modern animal rights movements advocate for the deconstruction of the human/animal dichotomy, challenging speciesism, and recognizing animals as inherently valuable beings (Regan 2003; Singer 1990; Francione 1996 in Freeman 2010). The term 'vegetarian' describes a diet which excludes all forms of animal flesh, while decisions whether to consume meat or not are usually ingrained within one's philosophical, ethical or religious beliefs (Kheel 2004). DeSolier (2013) signifies that in post-industrial societies the morality of production is immense; it has power over (re)shaping consumption. One's moral deliberation on what they ingest is ethical consumption.

The political importance of meat non-consumption is not possible without acknowledging the inherent political matter of animals themselves. Overall (2012) argues that whether an individual, human or non-human animal, counts as a person is partially dependent on how they are treated by others. An individual's significance derives not only from who they are, but also from how they are being engaged by others and how they engage with others. Considerations of multidirectional socio-political engagements should thus be central to all projects of feminisms.

Ecofeminisms are subfields of feminisms, which introduce and problematize the equivalencies of human and non-human animals being exposed to oppression. But valid considerations that should be taken are how far do ecofeminisms reach in regards to not abiding to identity politics, whilst aiming to be ethically and politically considerate of (every)one's significance?

Ecofeminisms emerged from collisions of cultural feminists, advocates of animal liberation movements, and anti-violence and/or peace protestations. The oppression of non-human animals, known as speciesism, is central to projects of ecofeminist analysis (Gaard 2002). Ecofeminism's central argument is that speciesism functions like other forms of oppression such as racism, sexism, classism, heterosexism, and naturism (Gaard 2002). However, amidst ecofeminists' indications that speciesism, sexism, racism, and homophobia must be considered and appropriated into analysis (Kemmerer 2011), some ecofeminists' discourses regarding gender are discussed solely on the level of the cisnormative gender binary system. This chapter will show how considering gender as an analytical category enables transcending limiting binary gender systems.

Ecofeminist theories arose from debates on nature among radical and cultural feminists in the 1970s and 1980s, indicates Sandilands (1999). Radical feminism argues for a feminist approach which centralizes the interrelations of sex and gender, depicting these interrelations as women's oppression by men (Sandilands 1999). Radical feminisms' conceive gender as cisnormative; one's gender identity is expected to be in accordance with their designated sex at birth, resulting in their argumentation deploying a reiteration of the gender binary system logic. Koyama (2006) claims radical feminism holds the belief "that women's oppression is the most pervasive, extreme and fundamental of all social inequalities, regardless of race, class, nationality, and other factors." Such rankings of oppression and simplistic usages of identity politics are inherently oppressive to people who are marginalized due to their intersections of multiple differential factors/categories, adds Koyama (2006). In accordance with Koyama, I am arguing that canons of vegetarian ecofeminisms delineate their argumentation for vegetarianism while manifesting the limiting usage of gender identity politics.

Vegetarian ecofeminisms are an active form of the feminist notion 'the personal is political' which explore and re-act to political contexts with dietary choices (Gaard 2002). They argue that only by acknowledging and acting upon our sympathies for animals, humans are able to prevent immense amounts of animal suffering, claims Gaard (2002). But, like all emotions, one's ability to be ethical and sympathetic is encaptured within social and political contexts (Gaard 2002). Vegetarian ecofeminisms argumentations thus seem to somehow not comprehend everyone's multifold encapturement in socio-politically (constructed) contexts. Let me elaborate; meat non-consumption involves material manifestations of contestations toward oppressive socio-political hegemonies. Within Western contexts, individuals therefore consciously make decisions about their food consumption, and manifest their decisions with actions fueled by their ethical and political values. So far vegetarian ecofeminists would agree. Where the logic of vegetarian ecofeminisms' argumentations seems to plunge into self-contradictions is throughout their conceptualization(s) of gender(s). Whilst seemingly critical of various hegemonic social categories, they seem not to engage in critical analysis of notions of genders. Rather, their argumentations appropriate everyone's gender as being in congruence with their sex desginated at birth.

The Sexual Politics of Meat was first published in 1990 and re-published with an additional preface in 2010, for its 20th anniversary. Adams is considered one of the most significant theorists of vegetarian ecofeminisms (Gaard 2002). The book is seen as one of the most prominent works of vegetarian ecofeminisms and resulted in serving as a primary source of reference for a canon of vegetarian ecofeminist scholarship (Donovan 1995; Lucas 2005; Donovan and Adams 2007; Nath 2011; Kemmerer 2011). Consequently, various works of vegetarian ecofeminisms appropriated Adams's arguments; my analysis of her text can thus be expanded to a canon of vegetarian ecofeminisms.

Permeated by emotional charge, Adams's work appropriates arguments for vegetarianism by presenting animal-derived food, meat in particular, as gendered female, and claims their consumption is a form of reinforcing oppression toward (cisgender) women and animals.[3] Adams's theoretical stance is positioned between feminism and animal liberation theories, and considers connections between sexism and speciesism through portrayals of how women and animals are conceptualized in the Western culture (Gaard 2002). Adams argues that important connections exist not only between women and nature, but also between women and animals, and argues for a feminist theory and practice which includes ecological perspectives in accordance with animals' well-being (Gaard 2002). Adams (2013) posits meat as a food item indicating power symbolism: "Meat eating measures individual and social virility" (48). She claims that dietary habits proclaim patriarchal distinctions: "Women, second-class citizens, are more likely to eat what are considered to be second-class foods in patriarchal culture: vegetables, fruits, and grains rather than meat" (48). For Adams, meat consumption is a male activity which manifests sexism, and the removal of meat is a threat to the structure of the patriarchal culture.

Adams directly correlates meat and dominance: "The male prerogative to eat meat is an external, observable activity implicitly reflecting a recurring fact: meat is a symbol of male dominance" (56). She links gender inequality with species inequality; obtaining meat and its consumption was primarily a male performance (58). Women, she continues, were/are gatherers and consumers of vegetable foods, which are associated with passivity, femininity (60). Adams's statements lead me to question her appliance of gender as an analytical category to argumentation. Adams's argument does not acknowledge the fact that if meat non-consumption is seen as powerful enough to fuel resistance to patriarchy, it can also contribute to analysis and de/re/construction of all notions of genders, not only those bound by the cisnormative binary. Adams's claims, arising from work that was "charged with gender essentialism" (Stange 1995, 18 in Gaard 2011, 34), remain within the gender binary system and do not acknowledge the agency that meat non-consumption has in correlation to oppression stemming beyond the patriarchal matrix.

Animals and women are made to be absent referents; this allows forgetting their existence as individual entities, claims Adams (2013). She depicts the absent referents as following:

> Metaphorically, the absent referent can be anything whose original meaning is undercut as it is absorbed into a different hierarchy or meaning; in this case the original meaning of animals' fate is absorbed into a human-centered hierarchy.
>
> (67)

Linking (sexual) violence upon women and violence upon animals within processes of meat production, Adams highlights how men dismember and objectify women and animals for their enjoyment and consumption (2013). She claims the values of patriarchy become institutionalized through the structure of the absent referent. Adams's absent referents has developed from within feminist stances, yet solely appropriating arguments to cisnormative notions of gender implies a construction of *other* women (e.g., transgender women) taking place. I suggest *othering* cisgender non-compliant individuals consequently disables one's meat non-consumption from consideration as a resource for critiquing gender identity politics and unacknowleges the political contentions toward patriarchy of transgender and cisgender non-conforming individuals' meat non-consumption.

Twigg (1986 in Nath 2011) argues that abstaining from meat is a rejection of dominant masculinity. Adams claims meat removal is hazardous to the structure of the patriarchal culture (2013, 62–63), adding that "men who become vegetarians challenge an essential part of the masculine role." Again, I question: What do such arguments imply about Adams's and vegetarian ecofeminism's comprehensions of gender(s)? Due to Adams's self-proclaimed positioning within the "radical feminist arena" (Tyler 2006, 123), my questionings in regards to her argumentations should not come as a surprise. Reducing genders to essentialist dichotomous notions, as Adams deploys in her arguments, disables analytical consideration of any form of gender identification except for those constructed as cisnormative. With forwarding transgender as an analytical category, I will aim to expand the ethical potentials of her arguments.

Transgender as an analytical category

While transgender individuals' fluidity and subversive embodiments contest hegemonic discourses of fixed identity categories, vegetarian ecofeminisms engage with identities as relationalities solely between cisgender men and women. Adams (2013) indicates the role of meat non-consuming embodiments as contesting hegemonic masculinity:

> Perhaps women's meaning is spoken in a different way at that point when they find themselves muted. Is it possible that food becomes the spoken language of dissent? Since women are the main preparers of food in Western culture and meat is defined as men's food, vegetarianism may carry meaning within a female language which seeks to escape its own mutedness.
>
> (213)

Juxtaposing diet and gender as analytical categories within vegetarian ecofeminisms would enable expanding such argumentation to the disidentifications of all other non-hegemonic genders whose meat non-consuming embodiments have agency. Confining argumentation within identity politics diminishes agency arising from material forms of contesting social hegemonies; for example, meat non-consumption. Claiming a man's vegetarian diet challenges instances of his masculinity and thus threatens patriarchy implies that men are essentially masculine. I question if argumentations based on forefronting females' experiences of oppression, which are instances Adams appropriates as enablers for women's ethical meat non-consumption, consider the possibility of female masculinity. And beyond this, would decoupling the linkage between cisgender women and animals, solidified within arguments that re-appropriate the patriarchal matrix, and transcending these linkages to social groups exposed to oppression due to various identificational categories, endanger vegetarian ecofeminisms' argumentation or add to its analytical depth and ethical and political solidarity? Ecofeminisms' queer critiques and feminist new materialisms will provide theoretical support for this elaboration.

Gaard's *Ecofeminism Revisited* (2011) depicts the downfall of ecofeminisms from a very promising theoretical and activist stance in the 1990s, to receiving harsh critiques and being avoided from scholars and activists by 2010. Due to over-intensely presenting women as embodied nature and cultural ecofeminisms' forefronting patriarchal notions, all ecofeminisms' were portrayed as essentialist by post-structuralists (Gaard 2011, 31). This can be related to ecofeminisms' tendency of conflating sex and gender categories. I am suggesting applying transgender as an analytical category to vegetarian ecofeminisms might be contributional to regaining their critical acclaim.

While ecofeminisms consider human and non-human animals central to their argumentation, their usage of gender as an analytical category seems to be onefold. Twine (2010) claims ecofeminists receive suspicions of essentializing identity categories, which supports my critique of readings of Adams's canon. Ecofeminisms as a movement links feminist and ecological struggles, claims Sandilands (1999), yet adds that these links need considerable unpacking, especially within their questionings of gender and nature. Sandilands elaborates her apprehension with forefronting that not only does nature have an important role in feminist discourses, but gender also has an important role in the social and political creation of the concept of nature (1999). Amidst indicating that both nature and gender work as discursive constructs, Sandilands exposes ecofeminisms' inadequacy with critically questioning the fundamentality and consequences these constructs have within discourse in Western contexts (1999). My readings of Adams's canon have produced similar observations, which can be attributed to the emergence of ecofeminisms from radical and/or cultural feminists.

Gaard (2010) suggests feminisms' theoretical legitimacy could be at risk, if there would only be advocacy for one group of females and ignorance toward intersections of their race, class, nationality, sexuality or species. I am adding gender to this list of individuals' intersections, and suggesting that vegetarian ecofeminisms withhold immense critical power with which the notions of genders can be analyzed and re-constructed to transcend exclusionary binary logic. Sandilands (1999) argues for more open and flexible explorations of identities and subjectivities, and places these processes within discourse, rather than denoting one's compulsory oppositional identificational factors prior to their presence in discourse. As I will now delineate, similar discourses in regards to notions of (gender) identifications and subjectivities have taken place within feminist new materialisms and transgender studies. I will connect those discourses with the notion of ethical meat non-consumption, upon which an ethically self-reflexive notion of transfeminism will develop.

The concept of nature is essential to (vegetarian) ecofeminists' theories. Some also echo essentialism—dichotomous divisions of *natural* and *cultural*. Kemmerer's (2011, 15) claim exemplifies dichotomies of vegetarian ecofeminists: "In Western patriarchal culture, both women and nonhuman nature have been devalued alongside their assumed opposites – men and civilization/culture." New materialisms intersect into these essentialist discourses. Feminist new materialisms emphasize the inclusions of "the biology of the body in cultural and political analysis" (Frost 2011, 69). Grosz (2008, 23) emphasizes that biology does not refer solely to "the *study* of life," but also to "the body, the organic processes or activities that are the *objects* of that study." Vegetarian ecofeminist analyses, especially those which appropriate radical feminism, are thus contested by new materialisms in regards to their usages of dichotomous divisions between nature and culture, sex and gender, body and mind.

Argumentations of Adams's canon are encaptured within binary logics of sex/gender, body/mind, nature/culture. Frost (2011, 76) claims that while feminists have been engaging in denaturalizing nature, they have not been "deculturalizing culture": assigning matter also to biological and material instances. Frost signifies that this very reluctance of feminists binds them to binary systems, which they are otherwise aiming to deconstruct. Exposing cisnormativity and usages of gender identity politics within Adams's canon provides grounds upon which I show that transcending binary logics can be only contributional and not detrimental to vegetarian ecofeminists' ethical and political solidarity.

Frost argues that within "calling for feminists to acknowledge that matter and biology are active in their own right, new materialists push feminists to relinquish the unidirectional model of causation in which *either* culture *or* biology is determinative" (2011, 71). Rather, feminists are exposed to demands of rethinking causation as complex and multilinear, and called to assess the possibilities for and toward socio-political transformations (Frost 2011). It is these calls that I am applying to Adams's canon, specifically their usages of gender identity politics amidst political projects of overcoming oppression, primarily speciesism, and cissexism. With the usage of transgender as an analytical category, cisnormativity of their argumentation in regards to gender(s) is depicted.

Transgender studies ask why does the fact that people experience and express their gender in fundamentally different ways ethically and morally matter, indicates Stryker (2006). The field also concerns itself with what can be done politically about the injustices and violence manifested upon individuals whose genders are non-normative and atypical (2006). Ethical values are thus of great importance for projects of transgender studies, in line with which I am arguing for the possibility of broadening vegetarian ecofeminisms' ethical projects by transcending usages of gender identity politics. Rather, openness and self-reflexivity is posited as central for ethical and politically solidary feminists' projects, building upon Stryker's (2006, 7) claim, that confronting topics of transgender studies "requires that some feminists re-examine, or perhaps examine for the first time, some of the exclusionary assumptions they embed within the fundamental conceptual underpinnings of feminism." Analytically critical feminisms, which I will present as ethical transfeminisms, consider transgender as an analytical category.

Conceptualizations of genders and their applications within research and analysis are means through which cisnormativity of vegetarian ecofeminism is being delineated. They are also a postulate upon which arguments for feminisms' capabilities to transgress gender identity politics and expand their ethical manifestations are applied through my argument. In *Imagining Transgender*, a critical ethnography of the category transgender, Valentine explicates transgender as an analytical category:

> I want to argue that "transgender," rather than being an index of marginality or "an out of the way category" (to paraphrase Tsing 1993) is in fact a central cultural site where meanings about gender and sexuality are being worked out.
>
> (2007, 14)

Valentine applies transgender to elaborate upon its political, theoretical, and ethical possibilities and limits. Rather than locating transgender at the margins of social intelligibilities of genders, transgender is being forwarded as a category which lies within the very constitution of gender as a social and analytical category. Transgender as an analytical category therefore has the capability of exposing how gender as a social category and hegemonic normative has been constituted. It is this capability of transgender that urges for its application and consideration to be included in analytical projects concerning genders and feminism.

Considerations of feminisms as ethical projects are thus inextricable from usages of gender as an analytical category. I am suggesting transgender be applied as an analytical category, which functions as committed to ethical and critical analysis. Transgender as an analytical category implies awareness and recognition of transgender identifications, yet does not analytically consider solely these identifications. Rather, transgender variance is emphasized, and accompanied with commitments not to abide constricting gender binary systems. Amidst contesting ultimate free rein for self-ascertaining of gender, Heyes (2003) indicates that one's gender expression might limit possible meanings and opportunities for others' genders. Ultimate individual freedom of gender expression would thus sidestep important ethico-political questions arising from gender relations and communal demands (Heyes 2003). My usage of transgender as an analytical category forefronts ethical considerations of others as an instance of gender expression, with aims not to hamper others' agency and social intelligibility within their gender expressions and identifications. This notion will be referred to as 'disidentification', inspired by José Esteban Muñoz.

"Disidentification is a performative mode of tactical recognition that various minoritarian subjects employ in an effort to resist the oppressive and normalizing discourse of dominant ideology," elaborates Muñoz (1997, 83). Disidentification offers individuals possibilities to reclaim and regain command of themselves within social spheres, without engaging in superiority of relations of excluding Others. Through disidentification resistance is enabled without mandatorily forming a counteridentity, it is rather a third term which resists the binary of identity and counteridentity. My usage of disidentification will follow Muñoz, yet also signify there are not always deliberate or primarily contentious processes of contesting mainstream ideologies taking place within individuals' performative acts. In other words, my usage of disidentification implies there is a possibility of the presence of transgressing social hegemonic ideologies within certain acts of performativity, however, there is no certainty of disidentification taking place. Rather, certainty of individuals' commitments to ethics can be traced within my application of the notion of disidentification. Resistance to normalizing and oppressive discourses is accompanied by ethical re-questioning of oneself in regards to others and with dedication not to take part in reiterating or re-creating exclusionary binary identities. Disidentification is therefore an instance of both matter and discourse, realms which are now to be entangled within my analysis and argumentations for ethical transfeminism.

Transfeminisms and new materialisms

Emi Koyama's "The Transfeminist Manifesto" introduces transfeminism as a movement primarily by and for trans women, and "open to other queers, intersex people, trans men, non-trans women, non-trans men and others [. . .] who consider their alliances with trans-women as essential for their own liberation" (2003, 244–245). Koyama explicitly notes that transfeminism has no intentions of overtaking existing feminist movements. Rather, transfeminism seeks to advance feminism as a whole, and do so by embodying feminist coalition politics within which each individual has the right to define their own identity and make decisions regarding their body. The notion of ethical transfeminism will build upon Koyama's conceptualization of transfeminism, whilst incorporating material notions of non-normative diets, embodiments, and gender disidentifications. Koyama's (2003) explicit indication that no one is completely free from the socio-cultural dynamics of the hegemonic gender system we all are embedded in will be continuously reflected upon. In other words, I will not present an argument which will call for formations of a utopian version of vegetarian ecofeminisms, but rather appropriate the cisnormativity of vegetarian ecofeminisms to exemplify how ethical self-reflexivity can be beneficial when analyzing and expanding incorporations of notions of genders, even though they are bound within hegemonic regulatory regimes.

Crucial political work of transfeminisms is their tackling the concept of privilege. Koyama claims that any individual, who has a "gender identity and/or an inclination toward a gender expression that match the sex attributed to her or him has a privilege of being non-trans" (2003, 246). Like other privileges, the privilege of being non-trans is invisible to those who possess it, while suffering due to its absence is severely familiar to those who are experiencing its lack (2003). Challenges of Adams's canon of vegetarian ecofeminisms' will be elaborated with regards to their un/acknowledgments of privileges, exposed based on their argumentations being solely in alignment with cisnormative gender concepts. I will follow Koyama' claims that essentializing gender identity is just as dangerous as omitting to biological essentialism in regards to the notion of sex (2003), and such work is not in alignment with transfeminism. Projects of transfeminism dismantle the essentialist assumptions of normativity of the sex/gender congruence and call for challenging ways through which socio-political factors influence our gender expressions and embodiments (2003). Transfeminisms thus contest the *naturalness* of biological sex or biologically sexed bodies, which is in agreement with conceptualizations of feminist new materialisms.

Manifesting contentions toward speciesism is central to politics of projects of vegetarian ecofeminism. As diets are political, their contestations take form through meat/animal byproduct non-consumption. Vegetarian ecofeminists are very meticulous and sympathetic in acknowledging the objectification and suffering animals are exposed to, and it is through these very "sympathetic connections" (Gaard 2002, 120) that self-reflexivity with regards to one's own stance toward oppression takes place. Vegetarian ecofeminists actually argue that it is not solely reason, but rather the combination of sympathy and critical analysis of cultural and political contexts which provide a reliable guide to ethics and their manifestations, elaborates Gaard (2002). Since their projects emerged, vegetarian ecofeminists have recognized the conceptual and structural similarities among sexism, speciesism and racism. With these indications in mind, I then question how has the notion of genders beyond cisnormativity seemed to bypass Adams's canon of vegetarian ecofeminists analyses and ethical self-reflexivity?

Manifesting expectations of acknowledging all intersecting personal identifications and categories are challenging to achieve. I suggest that the notion of ethical self-reflexivity, at minimum, calls for self-critical and self-reflexive theoretical analysis of all notions and concepts which are included in a certain analysis. Sandilands claims that women in movements of environmental justice and ecofeminism have not made issues of gender central to their political practices (1999). Gaard's (2002, 128) depiction of "feminism's commitment to inclusiveness" suggests that broadening inclusivity, and subsequent theoretical transformations, take place through recognitions of groups or individuals, who are being considered as socially unintelligible. However, a critique of eco-feminisms that Gaard (2011) considers as legitimately grounded was the exposing of essentialism of their notions.

There is a presence of gender essentialism within Adams's canon. Nowhere in *The Sexual Politics of Meat* (2013) does Adams analytically interrogate the concept of binary gender, within which her arguments are encaptured. When explicating what the sexual politics of meat refer to, Adams (2013) states:

> What is "the sexual politics of meat"? It is an attitude and action that animalizes women and sexualizes and feminizes animals. [. . .] *The Sexual Politics of Meat* is also the assumption that *men need meat*, have the right to meat, and that meat eating is a male activity associated with virility.
>
> (4)

The book's omnipresence of dichotomously dividing humans among women and men does not enter the discourse of questioning this division's construction. Rather, Adams's statements, such as "Men who decide to eschew meat eating are deemed effeminate; failure of men to eat meat announces that they are not masculine" (57), "One's maleness is reassured by the food one eats" (58), "Men who become vegetarians challenge an essential part of the masculine role" (63), are supportive of her essentialistic appropriations of gender identity to sex (gender designated at birth). This could be explained by the fact that Adams situates herself within radical feminism (Tyler 2006). Queer eco-criticisms, new materialisms, and transgender studies all contest essentialistic dichotomous notions of body/mind, nature/culture, sex/identity, which are appropriated to the cisnormative binary gender system. They do so by engaging in analyses which argues for applications of multiple material and discourse analytical categories.

I will now delineate possibilities for increasing the realm to which one's ethical self-reflexivity is applied by juxtaposing transgender as an analytical category and Adams's concept of absent referent. It is through such an increase that vegetarian ecofeminisms can transcend rooting their argumentation in cisgender normativity, and apply ethical commitments to contest gender identity politics. Transgender, when applied as an analytical category, highlights how formations and meanings about gender and sexuality are developing, claims Valentine (2007). Adams's conceptualizations of developing a feminist-vegetarian critical theory indicate that overlapping experiences of oppression render women and animals absent referents (2013). An absent referent allows forgetting about a human or non-human animal as an independent entity. When becoming an absent referent, that humans' or non-human animals' "fate is transmuted into a metaphor for someone else's existence" (2013, 66–67). Adams explicates the concept of absent referents as following:

> The absent referent is both there and not there. It is there through inference, but its meaningfulness reflect only upon what it refers to because the originating, literal, experience that contributes the meaning is not there. We fail to accord this absent referent its own existence.
>
> (67)

Being considered an absent referent therefore indicates one's original existence is no longer of value, or is only of value to indicate the value of hierarchically higher positioned beings. Animals, for instance, become meaningless, only providing meaning when considered within an anthropocentric hierarchy—giving value to humans, who consider consuming animal flesh as an indication of value, wealth, higher social positioning, health, and, in accordance with Adams, hegemonic masculinity.

How do Adams's deployments of cisnormative gender notions hamper her canon's manifestations of ethics? I am suggesting that individuals, who are not compliant with cisgender normatives, can be considered as being indicative of becoming an absent referent within Adams's canon. For one to become an absent referent, they must first be present. Animals become absent referents through butchering, and women become absent referents through cultural violence, rape, in particular, claims Adams (2013). Transgender/cisgender non-conforming individuals, who are neither mentioned nor discussed in Adams's canon, can thus solely indicate their stance would be that of an absent referent. Transgender individuals are not present—and so become absent. However, such absence, which stems from cisnormativity of Adams's canon, hints to Butler's claim, that oppression can also work covertly" "Oppression works through the production of a domain of unthinkability and unnameability" (2004, 126). She indicates that being unthinkable within a realm that "regulates the real and the nameable" also disables one's possibilities of disrupting political contexts (126–127). I am thus suggesting that appropriating argumentations solely to the cisnormative gender binary system, as Adams's canon partakes in, contributes to transgender individuals being conceived as socially unintelligible. Transgender individuals are thus not absent referents within vegetarian ecofeminisms, yet their absence still has implications for vegetarian ecofeminists' ethical and political commitments. "Thus, normative social spaces are structured around the presumed absence of disabled, queer, trans, and other marginalized subjects," claims Enke (2012, 75). Applying transgender as an analytical category to vegetarian ecofeminisms discourse could contribute to transcending these implications of invisibility and enhance ethical and political solidarity of Adams's canon. This very invisibility consequently also disables transgender individuals' meat non-consumption to be considered as agency of ethics within vegetarian ecofeminisms. Adams's appropriations of meat non-consumption solely as a contestation of patriarchy, thus present merely a limited array of political implications arising from meat non-consumption.

Potential for re-formations of vegetarian ecofeminist theories may therefore be hidden within approaching essentialist gender notions with usages of ethical self-reflexivity in juxtaposition with transgender as an analytical category. Agency of transgender individuals' meat non-consumption is an exemplifier of the possibility of ethical feminist projects to constantly be in a mode of re-constituting their theories and of re-questioning the scope of their ethical self-reflexivity. This notion can be contributional to re-constructions of feminist projects: distancing themselves from essential notions of gender identities and rather than reiterating concepts of gender identity politics, providing political work permeated with commitments to ethics and solidarity when re-considering genders. This will enable transgressing essentialist cisnormative notions of gender with interactional ethical self-reflexivity arising from considering meat non-consumption as agency of ethics. Vegetarian ecofeminist ethics, which are already being applied to considerations of non-human animals and cisgender women, could thus be transferred to an array of diverse disidentifications, creating a visceral form of ethical feminism: ethical transfeminism.

Agency is inseparable from relationality; no individual can view themselves solely as an entity without regards to intra-relations with and among others. Meat non-consumption as agency enables individuals to develop ethical self-reflexivity with regards to acknowledging others' experiences of injustice and oppression. Adams's canon of vegetarian ecofeminisms is ethically self-reflexive in regards to acknowledging the mutuality of (cisgender)-human and non-human animals' experiences of oppression. Their contestations toward oppression are manifested by ethical meat non-consumption—diet is thus a resource for political work. I argued for juxtaposing diet and transgender as analytical categories, which both have immense power of exposing oppression arising from socially constructed hegemonies and normativities. Rather than perpetuating discourse of cisnormative binary gender notions, vegetarian ecofeminisms can deploy ethical self-reflexivity to re-thinking the notion of genders, and thus transcend usages of gender identity politics. This would enable agency of transgender individuals' meat non-consumption to be contributional to enhancing vegetarian ecofeminisms' ethics when considering genders.

Transgender embodiments must be understood as having agency, for the possibility of arguing for their ethical and political impacts on vegetarian ecofeminisms. Various feminist delineations of the categories of sex and gender have provided a dichotomy between the physical signifiers of sexual attributes and identificational notions of gender. West and Zimmerman (1987) indicate people do not expect discrepancies between individuals' biological sex and gender presentations. Rather, they assume gender reflects biologically sexed bodies, which is in alignment with arguments of Adams's canon. This is inevitable; the canon is appropriated to the cisnormative gender binary system, which considers the dichotomy between sex as mandatory signifiers of one's gender, and gender identity as the compulsory identity attributed to a biological sex as inextricable. Adams discusses vegetarianism and binary notions of bodies as follows:

> If the body becomes a special focus for women's struggle for freedom then what is ingested is a logical initial locus for announcing one's independence. Refusing the male order in food, women practiced the theory of feminism through their bodies and their choice of vegetarianism.
>
> (2013, 213)

While bodies are thus seen as having agency to manifest critiques of patriarchy, Adams's argumentation does not apply ethical self-reflexivity to analyzing genders. I suggest applying transgender as an analytical category can contribute to re-thinking genders within discourse. Transgender studies would disagree.

Noble (in Enke, 2012) explicitly notes that *trans* bodies have always been present in feminism, even if these bodies have been ghosted by a belief that their presence has not been a part of feminisms or women's studies. Noble's statement correlates with Nagoshi's following elaboration. Feminist theories and analyses have widely questioned gender as a social construct, and its attributing of gender roles, indicates Nagoshi, Brzuzy, and Terrell (2012), yet points out that questioning of embodied female and male identity binaries has not undergone the same scrutiny. This means that concepts, which deploy cisgender identities as inextricably placed within the sexed body they appropriate, have not yet been broadly contested. Noble (2012) indicates that processes of *trans* illiteracy take place, and elaborates these processes as folding *trans* entities into non-critical binarized sex systems, which are only capable of making sense of bodies that are viewed as either female or male.

Central to projects of new materialisms are contestations of the naturalization of the body/mind and sex/gender dichotomies, which were widely embraced within various feminists' argumentation for women's equality—vegetarian ecofeminisms being among them. New materialisms forefront the notion that material bodies and their representations are not separate, and rather suggest there are bi-directional causalities between culture and nature/matter (Grosz 1994 in Lane 2009).[4] Within new materialisms, matter and the body are considered not only as they are constituted by language, discourse, and politics but also as being constitutive, having a distinctive kind of agency, claims Frost (2011). Meat non-consumption can be considered as agential also within Adams's canon, but solely within the realm of patriarchy/cisgender normativities. "Women may code their criticism of the prevailing world order in the choice of female-identified foods. In this case, women's bodies become the texts upon which they inscribed their dissent through vegetarianism," claims Adams (2013, 213). Due to agency of transgender meat non-consuming embodiments, they can contribute to disrupting feminist project enclosed into gender binary systems.

Applying notions of new materialisms to Butler's conceptualization suggests that both sex and gender are not merely given or assigned to the body, but are rather material representations of both cultural and biological power forces. While Butler locates all production of materialization and deployment of sex and gender normativities in cultural and linguistic discourse, new materialisms signify the inseparable and non-hierarchal intertwining of nature and culture. Participants' cisgender non-conformity and meat non-consumption can be seen as political progressive instances of bodies' materiality and matter. Bodies' agency thus stems beyond limits of discourse. This stance supports that transgender individuals' meat non-consumption, acts of producing bodies' as ethical and political materialities, can be contributional in ethical re-configurations of vegetarian ecofeminisms. Ethics of feminisms' theorizing transgender need to recognize the discursive limits in regards to individuals' self-transformations and aim not to deny agency to gendered subjects, claims Heyes (2003). Transgender meat non-consuming embodiments, constituted both of discursive productions and material bodies' agency, can be influential in contesting gender identity politics within vegetarian ecofeminisms, and offer a source upon which their ethical commitments can be expanded.

Transgender studies and new materialisms contest conceptual dichotomies— present in social constructionist arguments of vegetarian ecofeminisms. Following transgender studies and new materialisms, usages of identity politics within vegetarian ecofeminist projects can re-create discursive binaries, and not consider complexities of embodied experiences within analysis. My analysis thus forefronts considerations of arrays of expressions, experiences and disidentifications. This focus on *being* calls for feminist projects to be contentious toward usages of identity politics within conceptualizations and argumentations, which have expectancies of unitary experiences of identity. Rather, I am arguing for feminists' analysis to be committed to ethical self-reflexivity with regards to notions from material and discursive realms, as such feminisms are more consistent with ethical values, rather than feminisms based on identity politics. Transgender individuals' meat non-consumption in juxtaposition with cisnormativity of vegetarian ecofeminisms exemplifies this argument.

Participants' narratives included their yearnings for future developments and achievements of genders and feminisms. Almost all participants explicitly said there should be no essentialism, regarding embodiments/identities. Drawing upon theoretical notions and participants' narratives I have shown and/or argued for, the concept of ethical transfeminism has developed. Ethical transfeminism incorporates and intertwines instances of discourse and materialities in order to contribute to feminist projects committed to ethical self-reflexivity and critical analysis. Intertwinings of matters, which are both discursive and materialist, show that feminisms need not enclose their political projects to notions of identity politics. Rather, receptiveness to re-formations of oneself and others are pursued, and these ethically oriented processes are not based on abiding to binary identitarian notions.

Conclusions

Ethical transfeminism posits ethics as the central postulate upon which transfeminist projects are constituted. The notion of ethics stems from one's manifestations of ethical self-reflexivity—exemplified by a diet of ethical meat non-consumption and its implications for meat non-consumers. Ethics are thus manifested in relation both to oneself and to other human and non-human others. Ethical transfeminism does not argue for individuals to be meat non-consumers. Rather, ethical meat non-consumption is an exemplification of how material instances can be of great importance for expanding feminist discursive notions based on ethical values, and overcoming feminist usages of identity politics. The exposing of cisnormativity of Adams's canon was also used as an example of how appropriating argumentations to identitarian notions and the gender binary system can only impede ethical commitments of feminist projects. Ethical self-reflexivity forefronts self-critique and self-analysis to be applied to feminists' considerations of genders.

Ethical transfeminism in no way intends to imply that other projects of feminisms are not in accordance with ethical values, but rather forward ethical self-reflexivity as central to feminists' political work, and highlight how juxtaposing materialities and discourse enhance ethics and solidarity. Openness to re-conceptualizations of identitarian notions are sought for feminist argumentations to be as ethically self-reflexive as possible. "*Transfeminists* confront our own privileges" notes Koyama (2006, 247). Ethical transfeminism argues for re-formations and re-appropriations of discursive notions to take place with regards to personal and communal well-being. Recognitions of individual and communal lived experiences are enabled through ethical transfeminism being constitutive of intertwining discursive and material instances. Material instances imply taking individuals' tangible embodied *doings* and *beings* into analytical and theoretical considerations.[5] As Enke claims, trans theories insist that we challenge a cultural logic that believes that "the physical body is a site of identic intelligibility" (Ginsburg 1996, 4 in Enke 2012, 75). The power of material instances and embodiments being considered as agency is a postulate upon which cisgender non-conforming gender performativities have potential to contribute to re-negotiations of discursive notions of gender concepts. It is through recognitions of genders and gender disidentifications as material and embodied agency that constricting identitarian gender notions are transgressed.

To sum up, ethical transfeminism argues for transgressing binary systems and usages of gender identity politics to enable enhancing ethical commitments within feminist projects. Recognitions of the importance of applying ethical self-reflexivity to comprehensions of personal socially non-intelligible categories are forefronted. Ethical self-reflexivity arises from any form of embodied agency of ethics: material manifestations of ethical being or doing, which aim to contest social hegemonies. Comprehensions of anyone's experiences of oppression can give fuel to contribute to formations of politically solidary ethical transfeminism, which are committed to being ethically self-reflexive and analytical within their political work. Within such contexts transcendence of the dichotomous gender system is possible. Complexities of intra-related social categories, and arrays of gender disidentifications, are exposed to ethical self-reflexivity and self-critical analysis, which enables ethical transfeminism to be contributional in re-negotiations of socio-cultural intelligibilities of genders. Ethical transfeminism therefore argues for correlations of discursive notions and lived embodied experiences to be constitutive of feminist concepts and political projects. Such intertwining enables overcoming the usages of essentialism and gender identity politics, and broadening ethical and political solidarity within feminisms.

Notes

1 Hereafter referred to as 'Adams's canon,' indicating vegetarian ecofeminist works by Adams and authors befitting her line of argumentation.
2 Hereafter referred to as 'transgender individuals,' yet including all arrays of gender identifications, embodiments, and expressions, not merely those whose identification as transgender, but rather all who contest and/or are not in alignment with socially hegemonic cisnormativity. Cisgender originates from the Latin prefix *cis-*, which means 'on the same side'. A cisgender person is someone whose gender identity is in accordance with social norms posited for either women or men within the hegemonic gender binary system. Cisnormativity refers to norms, which assume and prescribe that individual's gender identity are coherent with the sex they were assigned at birth.
3 When addressing meat non-consumption, I am referring to flesh of all animals killed for food consumption. As Adams (2013, 111) claims: "People who eat fishmeat and chickenmeat are not vegetarians; they are omnivores who do not eat red meat. Allowing those who are not vegetarians to call themselves vegetarians dismembers the word from its meanings and its history."
4 New materialisms challenge the fact that matter is being depicted as passive, with reasons to undo the oppositions between reason and passion, self and world, nature and culture. Biological instances are constituted to have active, transformative forces of and within themselves, and don't rely on culture to prescribe them power.
5 These instances can include diet, dis/ability, gender, sexuality, body weight, ethic and/or cultural expressions.

References

Adams, Carol J. 2013. *The Sexual Politics of Meat: A Feminist – Vegetarian Critical Theory*. New York: Bloomsbury.
Butler, Judith. 2004. "Imitation and Gender Insubordination." In *The Judith Butler Reader*, ed. S. Salih and J. Butler. Malden: Blackwell.
Cherry, Elizabeth. 2006. "Veganism as a Cultural Movement: A Relational Approach," *Social Movements Studies* 5(2): 155–170.

De Solier, Isabelle. 2013. *Food and the Self: Consumption, Production and Material Culture.* London: Bloomsbury.

Donovan, Josephine. 1995. *Animals and Women: Feminist Theoretical Explorations.* Durham: Duke University Press.

Donovan, Josephine and Carol Adams, eds. 2007. *The Feminist Care Tradition in Animal Ethics.* New York: Columbia University Press.

Enke, Anne F., ed. 2012. *Transfeminist Perspectives: In and Beyond Transgender and Gender Studies.* Philadelphia: Temple University Press.

Freeman, Carrie P. 2010. "Framing Animal Rights in the 'Go Veg' Campaigns of U.S. Animal Rights Organizations," *Society and Animals* 18: 163–182.

Frost, Samantha. 2011. "The Implications of the New Materialisms for Feminist Epistemology." In *Feminist Epistemology and the Philosophy of Science: Power in Knowledge*, ed. H. E. Grasswick. New York: Springer.

Gaard, Greta. 2002. "Vegetarian Ecofeminism: A Review Essay," *Journal of Women Studies* 23(3): 117–146.

——. 2010. "New Directions for Ecofeminism: Toward a More Feminist Ecocriticism," *Interdisciplinary Studies in Literature and Environment* 17(4): 643–665.

——. 2011. "Ecofeminism Revisited: Rejecting Essentialism and Re-Placing Species in a Material Feminist Envirnomentalism," *Feminist Formations* 23(2): 26–53.

Grosz, Elizabeth. 2008. "Darwin and Feminism: Preliminary Investigations for a Possible Alliance." In *Material Feminisms*, ed. S. Alaimo and S. Hekman. Bloomington: Indiana University Press.

Heyes, Cressida J. 2003. "Feminist Solidarity after Queer Theory: The Case of Transgender," *Signs* 28(4): 1093–1120.

Kemmerer, Lisa. 2011. *Sister Species: Women, Animals and Social Justice.* Chicago: University of Illinois Press.

Kheel, Marti. 2004. "Vegetarianism and Ecofeminism: Toppling Patriarchy with a Fork." In *Food for Thought: The Debate Over Eating Meat*, ed. Steve F. Sapontzis. Amherst: Prometheus Books.

Koyama, Emi. 2003. "The Transfeminist Manifesto." In *Catching a Wave: Reclaiming Feminism for the 21st Century*, ed. R. C. Dicker and A. Piepmeier. Lebanon: Northeastern University Press.

Koyama, Emi. 2006. "Whose Feminism is It Anyway? The Unspoken Racism of the Trans Inclusion Debate." In *The Transgender Studies Reader*, ed. S. Stryker and S. Whittle. New York and London: Routledge.

Lane, Ricki. 2009. "Trans as Bodily Becoming: Rethinking the Biological as Diversity, Not Dichotomy," *Hypatia* 24(3): 136–157.

Lucas, Sheri. 2005. "A Defense of the Feminist-Vegetarian Connection." *Hypatia* 20(1): 150–177.

Muñoz, Jose Esteban. 1997. "The White to Be Angry: Vaginal Davis's Terrorist Drag," *Social Text* 52/53 15(3–4): 80–103.

Nagoshi, Julie L., Stephanie Brzuzy, Heather K. Terrell. 2012. "Deconstructing the Complex Perceptions of Gender Roles, Gender Identity, and Sexual Orientation among Transgender Individuals," *Feminism & Psychology* 22(2): 405–422.

Nath, Jemal. 2011. "Gendered Fare? A Qualitative Investigation of Alternative Food and Masculinities," *Journal of Sociology* 47(3): 261–278.

Noble, Bobby. 2012. "Trans. Panic. Some Thoughts toward a Theory of Feminist Fundamentalism." In *Transfeminist Perspectives: In and Beyond Transgender and Gender Studies*, ed. A. F. Enke. Philadelphia: Temple University Press.

Overall, Christine. 2012. "Never Eat Anything with a Face: Ontology and Ethics," *Planning Theory* 4: 336–342.

Regan, Tom. 2003. *Animal Rights, Human Wrongs: An Introduction to Moral Philosophy.* Lanham: Rowman and Littlefield.

Sandilands, Catriona. 1999. *The Good Natured Feminist: Ecofeminism and the Quest for Democracy.* Minnesota: University of Minnesota Press.

Singer, Peter. 2013. *Animal Liberation: A New Ethics for Our Treatment of Animals.* New York: HarperCollins.

Stryker, Susan. 2006. "(De)Subjugated Knowledges: An Introduction to Transgender Studies." In *The Transgender Studies Reader*, ed. S. Stryker and S. Whittle. New York and London: Routledge.

Twine, Richard. 2010. "Intersectional Disgust? Animals and (Eco)Feminism," *Feminism & Psychology* 20(3): 397–406.

Tyler, Tom. 2006. "An Animal Manifesto: Gender, Identity, and Vegan-Feminism in the Twenty-First Century – An interview with Carol J. Adams," *Parallax* 12(1): 120–128.

Valentine, David. 2007. *Imagining Transgender: An Ethnography of a Category.* Durham: Duke University Press.

West, Candace and Don Zimmermann. 1987. "Doing Gender," *Gender & Society* 1: 125–151.

Part III

Constructing connections

5 The women–nature connection as a key element in the social construction of Western contemporary motherhood

Adriana Teodorescu

Having children may seem so natural that the very fact follows the grain of human civilization. On the contrary, motherhood is simultaneously a personal feminine dimension, as well as a socio-cultural dimension which has always been influenced by power discourses and disciplinary strategies. Like any other socially anchored dimension of human life, motherhood is always socio-culturally constructed (Berger and Luckmann 1966).

Nevertheless, there is also a salient constant in perceptions and representations of motherhood through time, a constant that has shaped, and, as we will see, continues to shape the cultural representations of womanhood in general: motherhood seen as a natural expression of women's dominant features: a purely biological feature, namely the reproductive capacity (fertility, procreation, maternal instinct), and a bio-cultural feature derived from the first, namely the relational identity (care for the others, self-sacrifice). This powerful naturalist construction of motherhood had engendered social and literary narratives and, in turn, they have contributed to the enhancement of this construction. Ecofeminism, starting with Françoise d'Eaubonne, who coined the term in 1974 (d'Eaubonne 1974), has intensively reflected upon the numerous bio-anthropological and socio-cultural connections between women and nature. Both elements have suffered patriarchal domination, exploitation, and domestication (Mitten and D'Amore Ch. 6, this volume). But there is no consensus (Warren 1987; Cuomo 1998; Carr 2011) between ecofeminist scholars concerning the number and the specific content of cultural benefits for women, benefits that stem from the alleged women–nature similarity.[1] If there are scholars that emphasize the renewed need of the mother-earth rhetoric (Shiva 2005), with an emphasis on woman as nature goddess (Sjoo and Mor 1987), there are also scholars who contend that embracing the women–nature connection may lead to an idealized figure of nature, reifying the position that women are irrational and intended for reproductive purposes only (Cronan Rose 1991).

In this context, it is interesting to think that biology, as a manifestation or law of nature, was perceived as oppressive by real women, forced to bear multiple pregnancies, outcast in the realm of domestic chores and duties, educated in the spirit of fragility and of the body which constitute her greatest asset, but which have lessened a woman's autonomous potential. Despite discourses and practices in virtually all eras, unlike animals, women attempted to control their pregnancy in terms of number, outlook, and significance (De Angelis 2007). In other words, women have made a great effort to control motherhood, to separate it as much as possible from natural laws and to disrupt the connection between women and nature, even before the rise of the birth control pill, of the abortion movement, or of the feminist movement.

What I seek to point out in this chapter, embracing a constructivist perspective, is that the contemporary Western construction of motherhood is still underpinned by a strong emphasis on the connection between women and nature, but that this connection has its own cultural specificity that makes it somehow new compared with older versions of the same connection. I argue it is of utmost importance to acknowledge that the contemporary configuration of the women–nature connection plays the central role in the dissolution of motherhood ambivalence. By analyzing some of the social and cultural mechanisms that are building the submerged logic of this connection, I strive to underline the fact that envisioning nature in a positive light entails not only a naturalist, but also an idealized construction of motherhood. After a rather theoretical section, I explore two specific occurrences of the contemporary women–nature relationship within the discourse of what can be defined as the 'good mother' paradigm.

The 'good mother' paradigm and its social and cultural mechanisms

Postmodernity, and its inner relativity, leaves room for the idea that there is a plurality of experiences and feminine ideals (Hakim 2000), a fact which stands out in the conceptual foundation of post-feminism (Faludi 2006). However, motherhood is more uniform in what concerns socio-cultural representation, working as a diminishment of feminine differences. More than ever, motherhood is a value in itself for contemporary Western society, not only in what concerns its socio-economic importance (providing labor force and national prevail), but also in what concerns woman's cultural perception as mother. Motherhood tends to be viewed as a necessary stage in a woman's life that may be subject to delay but which should not be optional, no matter if it completes other dimensions of personality or if it represents the ultimate accomplishment (Jong 2010). Popular culture praises motherhood as a stereotypical, sugary display of affection toward an angel-like child through various media—films, news articles, women's magazines, books about child raising. The child is more important than the mother. As a result, while the father is more or less in the shadows, the mother consents to giving up her job (Brown 2004, 239) to raise and educate her child following closely the advice of specialists in child rearing. She embraces the natural birth movement (Brown 2004), opting for a vaginal birth, and, for a very long time, breastfeeds the child upon request, in accordance with the attachment parenting ideology (Warner 2006) and the strong advice of La Leche League International (Badinter 2012).

This unprecedented (Thurer 1995) obsessive focus on children, on meeting their needs, together with mother's desire to continuously improve as a mother based on an idealized perception of motherhood gave birth to a social phenomenon that scholars often place under the name of good mother/mothering[2] (Gotlib 2010; Thurer 1995), but alternatively employ also other terminologies such as *intensive mothering* (Hays 1998), *new momism* (Douglas and Michaels 2004) or *extreme parenting* (Dunnewold 2007).

A double and intertwined lack of ambivalence: motherhood and nature

The social imagery of the good mother paradigm displays a strong tendency to cast out any sort of ambivalence[3] from the way motherhood is represented (Brown 2010). At least a partial powerful explanation for this is the active presence—triggering new representations, not just ornamentally reiterating the old ones—of the connection established between women and nature, a nature imbued with highly positive, post-evolutionistic meanings.

Post-evolutionism refers to the unilateral way of perceiving and representing nature as an instrument which spreads life and as an endless reservoir of meaning. This vision on nature is specific to the culture of the last decades, tributary to a broken, spiritualizing reality, tributary to the moralist underpinnings of Darwinian scientific theory. Darwinian theory discusses the biological similarity between the human being and the animal, and evolution seen as a process of both perpetuating and differentiating the genetic endowment of mankind, a process governed by natural laws.[4] Yet, while sociologists like Norbert Elias underline that nature should not be forced to *make sense* and should not be considered intentional, on the basis of a tradition which endows nature with maternal symbolisms (Elias 2001), and while famous Darwinists like Richard Dawkins discuss the fact that the natural world contains no *rhyme*, no *reason*, no *justice*, while nature is claimed to be driven by indifference (Dawkins 1995), popular culture nurtures its need for meaning and order, in an unbothered way. Natural evolution is a kind of eternal celebration of life, entailing the totally necessary presence of death in evolutionism. Moreover, the post-evolutionist perception of contemporary popular culture associates nature with morality, and nature becomes, at an essential level, at least, good and fair. When nature is correlated with motherhood, nature becomes implicitly good, as will be seen from our second case study. Post-evolutionist nature suffers from an intense process of mythical transformation.

Indeed, the fundamental trait of the good mother paradigm is the glorious revival of the women–nature connection in the light of a post-evolutionistic grasp on nature, while women become, through childbirth and mothering, the agents that restore the ideological dominance of nature over culture. The post-evolutionist nature, which has lost its ambivalence, contributes to the dissolution of contemporary motherhood ambivalence. That is to be observed in several instances.

Raising an idealized child

In our contemporary Western era, ugly, nasty children are to be seen only in movies or artistic productions which deconstruct the classical image of motherhood and childhood. The cultural history of the meanings attached to children, that were not always so bright, is forgotten (Ariès 1965). The child is seen as a perfect product of natural laws, their most sheer biological expression. In this context, it becomes clear that a mother must always have good feelings toward her child. Motherly love is supreme love and stems from within, it is an instinct, a natural fact, which the woman experiences since pregnancy and which enables her self-sacrificial spirit. Also, because of the child's perfection, motherly love brings joy and happiness that helps her to cope with what Hays (1998) described as the cultural contradiction[5] between motherhood and femininity.

By rediscovering Liedloff's attachment parenting theories (1986) which are based on an evolutionistic background, the good mother encourages and cultivates the constant physical contact with the child, from birth onwards, when the child willingly leaves his parents' bed. She understands that eliminating the social expectations and rigor applied to the small child represents the natural and the best way to have a happy and mentally healthy child, a child for which the mother and not the school or any other institution is primarily responsible. The subsequent idea is that imitating the natural, animal, and especially primatological motherhood is the best thing to do concerning child rearing.[6]

Because children are perceived as part of nature, and nature is the equivalent of life, the good mother paradigm lays stress on a child's potential to ensure a parent with symbolic immortality (Lifton and Olson 1974; Van Tongeren 1995). Good mothers understand that their purpose in life is to have one or more children that will continue them after they will die. Moreover, the child is designed to offer society a carnal image of existence on the dark background of non-existence. Because of this, any child is better than no child, and having children at all costs is better than having none at all. However, the immortality substitute that a mother finds in her child is nothing but an illusion. According to Roy F. Baumeister, the medium duration that a child's symbolic immortality can offer a parent is seventy years (Baumeister 1991). Furthermore, the good mother paradigm seems to ignore the fact that through motherhood not only is life perpetuated, but also death (De Souza 2004).

Giving birth as the core moment of motherhood

By choosing to give birth, the woman resembles a godly figure, since she creates life out of nothing. The good mother paradigm insists on presenting the biological capacity of giving birth as a form of social empowerment, as an ontological advantage over men. What is tackled here is not just any kind of birth, but a specific type of birth, the natural, vaginal birth, which actually limits the number of surgical interventions involved: the epidural injections, the pain killers, and the C-section (Lamaze 1984; Odent 1994; Leboyer 2009). Women are advised to trust nature, because births have always happened in nature (a quantitative reason) and because nature is the mother which can take care of her own children—the environmentalist reason, which capitalizes on the traditional mother–nature metaphor (Hrdy 1999). On the contrary, women must reject medicine, a patriarchal science, which manipulates a woman's body, depriving mothers of the authentic experience of motherhood. One of the main arguments against the medically driven birth is the fact that it intervenes in the birth process, which is seen as the essential pillar of the mother–child bond. The more natural the bond, the better that bond remains.

The fact that pregnancy is a difficult time in a woman's life and that giving birth, no matter how, places women in close proximity to death is veiled in the good mother discourse by placing death entirely upon the shoulders of medicine. Thus, for mothers-to-be, fighting death implies fighting medical interventions.

The bad mother and the childless woman

What it is also characteristic for the good mother is that she knows to identify and sanction bad mothers. The bad mother (Lad-Taylor and Umansky 1998; Gotlib 2010) is egotistic, career-oriented and chooses everything that seems to go against nature: anaesthesis, C-section, even falsifying time of delivery, which leads to a traumatic separation of the child from the mother. She entrusts her child with a nanny, keeps the baby in a baby carriage instead of keeping the baby close to her body, and feeds the baby according to her own schedule, refusing to breastfeed for too long (if at all).

According to the good mother paradigm, the childless women must do everything they can to fix their condition. Those who are biologically able to have children must not postpone their pregnancy because it is not a natural gesture and the infertile ones must restore their natural ability to conceive, in the name of their natural-born wish to have children (Leyser 2010). Infertility is not considered natural. This is why medically assisted human reproduction becomes ethically justified. Leaving aside the economic reality that the fertility industry is quite prosperous, gathering around $3 billion each year, assisted human reproduction is solely based on the idea that nature is associated only with life, only with birth. Death, understood as non-existence which surrounds existence, is outcast from nature. The impossibility to give birth is deemed accidental and infertility is not considered to be nature's grasp on reproduction control. However, accidents would not be enough to justify the intervention of medical and pharmaceutical industry. An accident is, therefore, considered pathological.

The fertility industry overwhelmingly capitalizes on the female body (Le Breton 1990). The hormonal treatment involved in assisted human reproduction may be dangerous for a woman's health, in view of the side effects (multiple gestation, Ovarian Hyperstimulation Syndrome, ectopic pregnancy, ovarian cancer, breast tumors, etc.). Actually, medically assisted human reproduction sets the maternal sacrifice instinct against a consumerist background. In order to have the desired child, the woman must sacrifice herself, but her sacrifice is relative to the cost-benefit equation. Having a child will always be better than having none. A benefit which can also diminish a mother's sacrifice sensation is the sense of woman's empowerment. A woman may overcome her natural resources (e.g., her diminished ovarian function) and her reliance on her partner (e.g., when donor sperm is needed).

That the woman's freedom of choice is socially manipulated is disturbing and it signals the contradiction between the medical aspects involved in assisted human reproduction and the naturalization hailed by the Western good mother paradigm. Compared to the cultural mother–woman paradox (Hays 1998), this contradiction does not stem from the co-presence of non-maternal and maternal femininity. The contradiction actually stems from birth itself, the thin line before the moment birth takes place. Up to reaching this line, everything is permitted to have children, thus including the overuse of medical intervention, aggressive as it may be, risky as it may be, facts which are rarely pointed out in specialized literature (Hansen et al. 2013).

A woman's wish and desire to have children and also her pregnancy are subject to intensive medical care, while in the socio-cultural imaginary delivering babies is associated with the realm of the natural. The natural birth discourse highlights the advantages of natural birth for the babies who are healthier and more intelligent if born the natural way, while being prone to allergies, obesity, diabetes, or deficient immune system if C-section delivered. There is a wealth of studies vouching for this difference, while there are significantly fewer studies dealing with the health differences in naturally conceived children vs. those conceived via assisted human reproduction.

Breastfeeding images: the war against clothes, time tables and other social constructions

In recent years, motherhood has become a very popular theme for Western photographic albums. Both the pre-partum and the post-partum periods are subjects of interest for contemporary photography professionals. While, in the prenatal period, the images focus on the mother's large, exposed belly, often covered in colorful drawings of smiles and little eyes, indicative of the life residing therein as well as of the mother's happiness, in the postnatal period, the most frequent theme is breastfeeding. Events such as World Breastfeeding Week, celebrated during the first week of August in over 150 countries since 1992, overemphasize, especially in recent years, the role of the photographic products devoted to breastfeeding (Liao 2015; Bologna 2014).

I intend to discuss some of these very recent photographs which portray breastfeeding mothers, by focusing on the way in which the woman–nature connection contributes to the definition of the contemporary motherhood profile in the West. More exactly, I will refer to the photos taken by American Ivette Ivens and to *Women in the Wild*, the album of American Erin White (Erin White Photography 2015), who currently lives in Germany. White's album is still an ongoing project, as she is taking photographs in the USA, seeking, in the near future, to shoot photos of women breastfeeding in Europe (alone or in a group). Ivette Ivens gathered all her photographs of breastfeeding mothers in a book entitled *Breastfeeding Goddesses*. Some of the images which make up the book can be accessed on the artist's Facebook page (Ivens 2015) and in the papers (Ifeanyi 2015; Purcel 2015; Dainius 2015), which all praise the artist's project and initiative.

The rhetoric on which these breastfeeding images are constructed, as well as the texts which accompany them, is one meant to set this reality free from the artificial bonds of Western society, a society characterized by a massive repudiation of nature from the inter-human relations and by the rejection of what is understood as natural. This revolutionary rhetoric draws an alarm signal regarding the postmodern condition of motherhood—precarious at the level of public, inter-body means of expression (mother–child), subjected to censorship especially in the baby's first months of life—and a strong advice emphasizing the necessity of not only contesting, but also transcending the social norms which are restricting for the woman who becomes a mother. Beyond the aesthetic ambitions of the albums (the photography as work of art) and beyond their practical functions (the photography which serves the community wishing to capture motherhood snapshots), the major objective of these images is to raise awareness regarding the normality of breastfeeding, especially in the public space, in a world where, as Krista Zurn explains commenting on the album *Women in the Wild*, we are, as society, explicitly open to everything sexual and violent, but we maintain the taboo on breastfeeding as far as visual representations are concerned (Zurn 2015).

For both artists, nature is the main element which disintegrates contemporary bias at the level of mentality and imaginary. The underlying logic is that motherhood in general,[7] and breastfeeding especially, is a *natural* component of femininity. Therefore, exterior nature, the surrounding environment, little polluted by human intervention, and many times wild, becomes the best scenery for women who breastfeed alone or in a group, at any time, during the day or the night, on the brink of a waterfall, in a hammock, in a sunny meadow, on the sea shore or near a river bank behind which a city can be seen, while travelling, etc.

The major trait in these albums, in terms of both exterior nature and feminine nature, is that nature is a construct, first, because the natural, negative dimension of nature (nature is also equivalent to cruelty, ugliness, nurtures competitiveness instead of cooperation, there are mothers which abandon their offspring, etc.) is completely eliminated, and, second, because the authors resort to the glamorization of external settings and of the woman's feelings regarding her own motherhood. As can be noticed, following a brief examination of the photographs, all the natural environments are friendly, and some are even spectacular—red dawn, lavender fields, etc.—and all mothers, irrespective of whether they breastfeed one or more children, irrespective of whether they are accompanied by other women or not, are glowing with happiness, are overjoyed. There is nothing regarding the torment many women go through when breastfeeding, nothing regarding the conflict generated by the desire to breastfeed the child and the necessity or desire to have a life which does not include the baby, for instance, a professional life.

An interesting aspect which must be mentioned is that some photographs insist on the similarity between human motherhood and animal motherhood, establishing an evolutionary line and connection, between women and female animals in general.[8] For instance, in one of the photos by Ivette Ivens, a woman breastfeeds, on the ground, in an interior setting, a rustic room, similar to a chalet in the middle of nature. During this time, the woman pets a cat surrounded by her offspring. This similarity becomes more apparent as we can notice even in other photographic albums by Ivens: the tendency to gather toddlers and young animals side by side, the attraction toward the women who breastfeed one or two children of different ages, while being surrounded by even more children who are too big to be breastfed. The hidden narrative which can be identified here is one which fully de-culturalizes motherhood by positioning it not only in the middle of nature, but in its very animal center.

Coming back to the glamorization of external nature and of feminine nature, it can be noticed that the process is twofold, and that its directions many times overlap. One direction casts a romantic light on motherhood, while the other casts an erotic light on the woman. Romanticism simultaneously belongs to external nature and to the woman who exerts her mother role through breastfeeding. The clearest signs of romanticism feature the flower crowns and the leaves which adorn the braided hair or the lavish hair let loose (for Ivens), the long dresses, some blowing in the wind, the deep cleavage despite the winter outside (for Ivens), the women's intensely dreamy expressions. Nature, irrespective of how *wild*, functions, at the level of visual suggestion, as an immense cradle for the mother and the child. Nature is only a formal barrier at times, meant to better underline the symbiotic body of the mother and the child as in one of Ivette Ivens's photographs, where the woman is lying on the sand, surrounded by vegetation. Erin White shows preference for a diffuse, tender, soft light at dawn or filtered by the tree branches.

The second trend in the glamorization of nature is the erotic light shed on the woman. Eroticism should be understood as an intended overemphasis of the woman's erotic nature. Both artists want to heal the damage motherhood inflicts on feminine eroticism by resorting to nature, as if nature would offer even the benefit of re-sexualization, of rediscovering the erotic dimension. In most cases, the woman's body is unveiled, and not only in terms of the breast used to feed the baby. Both Ivens and White capture women dressed only in their underwear. Sometimes the underwear is not present, but full, frontal nudity is avoided in such cases. In some photographs women wear a scarf, a skirt or a dress, but there are other even less *likely* items of clothing in a realistic setting: a pullover over a completely naked body or a light summer dress in the winter season, clothes which, together with the makeup stand as proof for the artificial way of constructing nature. Eroticism first stems from the unexpected meeting between motherhood and the display of the body. Motherhood, breastfeeding in particular, presupposes the unveiling of the body. However, this does not automatically lead to the fact that every representation of motherhood should be based on unveiling the body. Yet, this is exactly what the two photographic albums promote. Moreover, there are mothers for whom the joy of breastfeeding is presented under the form of an ecstasy at the border between religiousness and eroticism, an orgasmic ecstasy which is best reflected by the facial expression: the gentle open mouth, the head bent backwards, etc.

The shape of a woman's body no longer necessarily respects the excessive standards of Western slenderness.[9] However, fat, cellulite, the natural body imperfections do not prevent the erotic filter cast over the woman's body. The greatest achievement of these albums is that, from the point of view of the feminine body, they give rise to a feminine, plural, non-restrictive eroticism, focusing on non-phallocratic eroticism. Men are very rarely present in the photographs belonging to the two photographers, being obvious that breastfeeding mainly concerns the mother and her children. When they do appear, men are mere witnesses of this biological wonder, glancing the perfect bond between mother and child with awe, as if a goddess would be born from the foamy waters of motherhood.

Up to a point, these albums sympathize with a war declaration against the social constructs of Western motherhood understood as a means of dominating the woman and the relations she entertains with nature through her motherhood. The benefits that any social deconstruction includes are present here as well: a benefit which is specific to the relativization of the patriarchal discourse of a motherhood hidden behind clothes and peripherals, if we think that, at the level of visibility, motherhood cannot yet compete with a woman's non-maternal eroticism—and the general benefit of training one's critical thinking regarding social reality, which is always capable of concealing the mechanisms via which it articulates itself as cultural construct. However, beyond this point, there is a risk that the war declaration which these photographic albums entail contributes in a large measure to *supporting* the happy motherhood ideology within the Western cultural paradigm of the good mother. This is because such albums promote a biological Messianic aspect: the woman (she is a *goddess*, according to Ivens) is excessively confronted with her reproductive and life support functions up to a point where their denial is impossible. Breastfeeding can no longer be discarded or rejected if we fully accept the cultural narratives of such photographic albums. Any form of disinterest in the face of motherhood is no longer an option. This point is even more valid since motherhood is associated, through nature, with eroticism, coming to terms, at least on the surface, with the cultural contradiction of motherhood. By returning to a nature, which is understood in perfect synonymy with normality and simplicity, but paradoxically, with a paradisiacal beauty, motherhood atones the woman, purifying her from the burden of Western civilization, rendering her available to a biological essentialism (there is a kernel of womanhood called motherhood), by its reductionist portrayal (motherhood is reduced to exerting biological functions, which, since they are natural, they are considered good and beautiful), and its generic, stereotypical depiction (all women breastfeed and all are thus happy).

Another problem can be noticed by analyzing the goal of these photographs. The photographs are meant to render public opinion more sensitive to the normality of breastfeeding, and implicitly to encourage the rise in the number of women who breastfeed in their private environment, and in the public settings, in particular. This sensitive way of approaching things is established on a principle of transitivity of consumption. To be more precise, what is targeted here is the fact that seeing images of women who breastfeed automatically leads to a greater acceptance of the presence of women who breastfeed as such, in non-artificial ways or settings. While some of the portrayed women can be easily identified through their natural traits, others look more like celebrities. This might prevent imitation, or, at least, make it less immediate or predictable. Moreover, the majority of the settings where women who breastfeed are placed, despite their pretence of being *natural* through the quasi-paradisiacal atmosphere they emanate, seem rather constructed, again risking the dissuasion of mothers-to-be from breastfeeding or, at least, from the manifestation of such a gesture in the public space characterized by socio-functional ambiguity and unpredictability. It is difficult to assess how many women will feel encouraged by these images[10]—as noticed after Erin White's campaign, when many women met the photographer, in various locations, to be photographed breastfeeding—and how many will assess the situation as a movie in which they are not prepared to play.

Cesarean awareness month: there is no cancer and no C-section in paradise

The fundamental ideological pillar of the "good mother" is natural birth and this remains a necessary moral imperative even when it is accepted and encouraged that women who cannot have children should resort to the procedures of reproductive medicine. Women thus tamper with a nature which is considered good, in principle (since it involves birth), and flawed in some circumstances (when the woman cannot remain pregnant or suffers from miscarriage). The more natural birth is wanted, the more is C-section perceived as a threat, being seen as the agent with the highest dissolution power over the motherhood universe. Moreover, while in the good mother paradigm natural birth is the general means of validating the mother's identity, the C-section calls this identity into question.

In what follows, I will analyze some of the manifestations and a portion of the socio-cultural goals of a phenomenon which took on a large media importance. It stirred the creation of an organization which leads an ample campaign against C-sections. The organization I refer to here is *The International Cesarean Awareness Network (ICAN)*, which perfectly integrates with the natural birth movement, an organization which defines itself as:

> a nonprofit organization whose mission is to improve maternal-child health by preventing unnecessary cesareans through education, providing support for cesarean recovery, and promoting Vaginal Birth After Cesarean (VBAC).

The organization admits to the fact that the C-section is necessary in some situations, but fails in rigorously delineating what 'necessary' means. However, it is clear that necessity can be calculated only considering a strict medical agenda: saving the mother and the child. ICAN anchors its campaign in the recommendations of the World Health Organization, which has repeatedly expressed its concern for the increase in the number of C-sections which are not medically necessary, in developed countries especially. The following statement from April 10, 2015 argues that:

> Since 1985, the international healthcare community has considered the "ideal rate" for caesarean sections to be between 10% and 15%. New studies reveal that when caesarean section rates rise toward 10% across a population, the number of maternal and newborn deaths decreases. But when the rate goes above 10%, there is no evidence that mortality rates improve.
>
> (World Health Organization 2015)

The mortality rates are the ones which are taken into consideration when the decision for C-section is made. Other criteria, like the pregnant woman's comfort, are not considered. Moreover, the potential negative effects of C-section are always mentioned, even the serious complications, like disability or death, which are more common in medical establishments which do not have the necessary facilities for surgery. If we consider that there are high chances that these establishments, incapable of offering infrastructure for surgery, are found in less developed countries, where C-section is not so common, a brief conclusion would be: the pain is here, but the wound is elsewhere. The problem is that the number of C-sections is on the rise in developed countries, even if the mortality rates are higher in less developed countries. A similar situation can be found in the case of the ICAN discourse: with the exception of a sound medical reason,[11] there should be no other type that would determine a woman to choose C-section.

It can be understood that, from their perspective, an emotional, affective choice of C-section, in agreement with personal criteria, to the detriment of natural birth would be a serious stray from the good mother model. The ICAN discourse and the many articles which support the Cesarean Awareness Month underline that C-section should be chosen for medical reasons only. Also, the organization stresses the differences between natural and C-section birth, so that the reader feels the profound conflict on which this discourse is built. Natural birth is safe medical-wise; natural social-wise, a chance for women to be actively involved in the child's arrival in the world, a unique moment that women will remember all their life. On the flipside, C-sections are fraught with medical risk, are not natural and are socially stigmatized. ICAN does not want endorse this stigma, but does not strive to criticizing it, as it is considered inevitable somehow. From a personal standpoint, C-section is traumatizing for the mother. ICAN uses the confessions of some mothers who went through C-section and who later became supporters of Vaginal Birth After Cesarean (VBAC), mothers which portray C-section in bleak terms:

> Many women who have had these unnecessary C-sections struggle to find support and outlets that help them deal with the sadness they feel over their birth. [. . .] Birth creates a new life as a tiny baby, but a mother is also born that day as well. She deserves to have her feelings validated. [. . .] I do not remember much about the operating room, just feelings and sounds: the clanging of metal instruments, banter of the staff, and wanting to move my arms which were spread out and strapped down. The hospital staff were the first people to see my daughter as she was pulled from my sliced abdomen. [. . .] I was the last person in the room to see my baby.
>
> (Jaclyn 2014)

> My scar still hurts/I was separated from my baby/I failed the one thing I'm supposed to be able to do as a woman.
>
> (Terreri 2015)

On the other hand, the organization does not want to lose the ones, who, at a certain point in their lives, opted for C-section, but wishes to persuade them not so much that natural birth is superior to C-section—as these can be easily understood—but that natural, vaginal birth can occur after a C-section. ICAN presents itself in an altruist light, as it is willing to forgive, understand, and offer support for the women who were led astray. Basically, these women make up a significant portion of the target segment of the ICAN organization and of its supporters (Lamaze International 2015). The supporters want to save the women who were socially stigmatized for giving birth through C-section, without manifesting their intention of contesting the stigmatization process. Women must be helped because C-section is something bad and they, the women, are only partially responsible, the main fault being attributed, through impersonalization, to the medical system (Vireday 2014; Muza 2015), to postmodernist society, or to anonymous factors.

The symbol of Cesarean Awareness Month is a dark red ribbon. The symbol is explained in the following way:

> The burgundy color of the ribbons represents birth and the wearing of the ribbon upside down symbolizes the state of distress many pregnant women find themselves in when their birthing options are limited. The loop of the inverted ribbon represents a pregnant belly and the tails are the arms of a woman outstretched in a cry for help.
>
> (International Cesarean Awareness Network 2015)

The fact that people talk about the limited options women have regarding birth is a euphemism for laying the blame on the existence of C-section as option. The problem of the organization is not that there are not too many options, but that C-section is one of these options. Here we get at the heart of the campaign ICAN runs: casting a pathological shadow on C-section. From their perspective, it can be understood that resorting to C-section, despite being a medical act, implies becoming ill, letting the body be filled with signs of disease. The symbol of cesarean awareness month, the dark red ribbon, is constructed in an analogy form with the pink ribbon used to symbolize breast cancer awareness. This dialog between symbols is very dangerous because it draws on the similarity between cancer and C-section too much, even when the differences are obvious. Cancer is not the result of a conscious choice. Cancer, unlike C-section, cannot save lives. There is a common element between the two, though. Both showcase the fragility of human beings seen as products of nature and the ambivalence of nature per se. On the basis of this common element, ICAN's imagery placed cancer and C-section side by side.

What is still to be discussed is the double, schizoid standard, applied by the contemporary society which praises the naturalist model of the good mother over the conception and birth of children. While, in terms of conceiving children, everything should be done to remedy infertility, through the possibilities of medicine, childbirth should be as natural as possible, since the medical community generates malicious comments, blame, and criticism. The risks the woman exposes herself to when resorting to medical assisted human reproduction, and the inherent risks of any natural childbirth are not set in the collective memory of society and contemporary mass culture. In stark contrast, the risks of C-section are promoted, debated, and fought against. Sometimes, when these risks are discussed, popular culture counteracts with another narrative of motherhood: the necessary human sacrifice of the mother. Birth pain is particularly considered an essential component of the sacrifice the mother must endure to bring a new life into the world. The fact that contemporary Western society maintains, on an imaginary level, the necessity of sacrifice cannot be solely explained as traditionalist reminiscence,[12] as it is largely determined by projecting motherhood onto a naturalist and evolutionist setting.

We must admit to the fact that, as in the case of photographic albums showing women who breastfeed, there is a benefit to this campaign. It draws attention toward using surgery to bring babies into the world as a phenomenon with numerous negative implications, contributing to a rise in women's awareness and to an encouragement in looking up information. But the benefits do not go beyond this awareness because the ICAN discourse is restrictive and manipulative. The discourse is restrictive because it wants to steer women toward the good path of natural birth. The discourse is manipulative because the method by which ICAN understands manipulation is hiding a general truth about all births. Any birth, whether natural, vaginal, or surgical implies a series of risks for both the mother and the child, the highest risk being death (World Health Organization 2015). The risks install during pregnancy.[13] Admitting to this fact would entail that nature offers no guarantee, no form of automatic superiority. As such, the woman cannot be obliged to accept the path nature offers, as it may not be the royal path of her social and personal destiny.

Whether in newspaper articles (Curtis 2014, Feinmann 2011) or blogs (Tuteur 2011), the Internet is filled with opinions which reflect the conflicting, manipulative rhetoric of ICAN, which advocates in favor of the moral and social superiority of vaginal birth. This idea of vaginal birth strengthens the good mother portrait, a mother biologically devoted to her child, contributing, by opposition, to the creation of the bad mother profile. For example, a young woman blogger argues against the fact that the birth story is so important in defining the maternal identity of women, and argues in favor of the private nature of birth, seen as the first, biological step of motherhood, and not as its essence. She is appalled by the ICAN campaign:

> Yes I do believe there are too many unnecessary C-sections in this country, but who am I to judge how another woman brings new life into this world? It's none of my business. It's also unfair that women who have medically necessary cesareans need to preface their birth with this fact so that they aren't judged [. . .] We are immediately judged and given the side eye. I'm done being judged and I'm ready to stand up for other women who are being judged and feeling bad about themselves because of it. No more. Yes I know it's dangerous, yes I know recovery is hard, yes I know that most people think the C-section rate in the US is too high, but you know what I don't care. And it isn't for anyone to judge how another woman became a mom. Also let's be real here, it doesn't matter how you give birth . . . the next week or two is hard regardless of how the baby came out. No one should ever be ashamed or embarrassed about their birth story. Whether you delivered your baby vaginally or by a cesarean congratulations on becoming a mom.
>
> (Growingupawife 2013)

Recently, Eliane Glaser (2015) noticed that natural birth functions as a positive stereotype in our contemporary society, and that women do not have a real possibility to choose between several birth procedures, being compelled by a strong anti-medical naturalist trend to embrace natural birth. She doubts the real empowerment engendered by the naturalist discourses, observing that the focus on the female body undermines women's capacity to reason:

> Natural childbirth has mistakenly come to be regarded as automatically woman-centred, with midwives portrayed as helping women—in the teeth

of the white-coated male establishment—to achieve the authentic experience they supposedly really want. But being bullied or cajoled into having a natural birth because of trumped-up risks to "baby" is not what I call feminism [. . .] It is right that a woman should have control over her own body. But in the cult of natural childbirth, the body is everything. Telling women to "just let go" can be liberating, but it's also a command to switch your brain off.

Conclusions: for a healthy farewell to motherhood essentialism—breaking the women–nature connection

The idea of having children might be the last essentialist value of the Western world. Researchers (Hernandez 2010) speak of the fact that idealizing motherhood, by using elements pertaining to a romantic view upon nature like feeding, caring and self-sacrifice, tends to compensate the sense of capitalism's economic instability. Ecology blends with cultural Christian heritage to engender a liminal space, at the border of reality and illusion: the heavenly realm of good motherhood. But revenge from the negative side of motherhood insinuates itself perfidiously. Mothers who give up their jobs in order to appeal to the requirements of the good mother do not *re-instate* a natural lost world, but instead manage to capitalize on and increase women's poverty, poverty being a corollary of women anyway (McLanahan and Kelly 2006). Moreover, the feeling of personal happiness diminishes (Baumeister 1991), the risk of depression grows (Evenson and Simon 2005; Rochman 2011) because of the responsibilities involved in raising children and the social pressure to be a perfect mother (Crittenden 2001). Let's not forget: in a good mother paradigm happiness is mandatory. But when real women cannot be good mothers, happiness falls apart, while frustration and blame settle in, highlighting the dark side of a naturalist construction of motherhood.

God is dead, said Nietzsche. As Foucault would say, Man is just a sand figure, soon dissolved by the sea. Marriage is an emotionally overcharged social configuration that sociologists warn is headed for extinction (Klinenberg 2013). The time of grand absolutes and essentials is long gone. Western society must prepare to say goodbye to motherhood essentialism. And for this it is necessary to acknowledge and to tear down the connection between women and a mythologized, perfect nature.

Notes

1 As Luca Valera (Ch 1, this volume) points out, the works of Françoise d'Eaubonne still have so much to offer to those interested in both nature and women's liberation, mainly because d'Eaubonne has the willingness (even if not always the power) to overcome the traditional dichotomies that characterize Western thought. These dichotomies (e.g., women–nature/men–culture) are expressions and triggers of patriarchal domination over women's bodies and sexuality as well as over nature.

2 When I refer to the *good mother*, I refer to an emblematic symbol of Western society, a rhetorical and ideological figure, not to the women who more or less approach this symbol.

3 Although ambivalence characterizes human reality. See Durand (1960).

4 More about the post-evolutionist perspective on nature, and for a critical approach on the subject, see Teodorescu (2012).
5 A woman's values, as praised by mass culture—eroticism, freedom, individuality, slim body, career—are rendered invalid by the cultural imperative of the good mother.
6 A very insightful analysis of how endorsing primatological narratives generates an essentialist interpretation of human motherhood is to be find in Sowards (2011).
7 *Attachment parenting* and those inspired by these trends/organizations do not restrict themselves to considering breastfeeding the only natural gesture, although they assign it special importance, focusing on the idea that breastfeeding fundamentally sets the bond between a mother and her child. For them, motherhood itself, as a whole, is the living proof that women belong to nature, and, conversely, that nature is written in feminine genes, transgressing the cultures in which women live.
8 For other contemporary representations of the analogy between women and animals and for an astute discussion of some of the major consequences concerning the social perception of gender roles, see the incisive study of Stephanie Baran, "Visual Patriarchy: PETA Advertising and the Commodification of Sexualized Bodies" (Chapter 3, this volume), who debunks the apparently well-intentioned advertisements that associates women with animals in order to fight for the animals' rights. Baran contends that insisting on the similarity between women and animals, more specifically using staged images of women as symbolic substitutes for animals' conditions reinforces the commodification of the female body, without being useful in really drawing attention on animals' issues.
9 For an eco-feminist critique of the Western standards of unrealistic feminine beauty, see Mitten and D'Amore (Ch 6, this volume).
10 Not every visual narrative that technically emphasizes the connection between women and nature becomes automatically successful in re-establishing a genuine encounter between women and nature. For an analysis of such a narrative, which seems to rather put distance between women and nature, see the study ofValerie Padilla Carroll, "Writing women into back-to-the-land: feminism, appropriation, and identity in the 1970s feminist magazine *Country Women*" (Ch. 7, this volume).
11 The organization and its supporters admit to the fact that, in many instances, C-section can save lives, but that the procedure is justified only when life is threatened. It is as if in such discourses, the collective imagery of surgery-driven birth is blocked in the Middle Ages, when, indeed, because of the lack of anaesthetic and professional instruments, C-section was an almost barbaric practice, performed on dead rather than living mothers. See Brown (2004).
12 Everywhere, in all the cultures, the mother's self-sacrifice is an important component of the imaginary domain and of motherhood's discourse. In the capitalist world which praises individualism and career, self-sacrifice remains a necessary baptism for the woman who wants to assume a mother-type function. This happens even if physical sacrifice becomes blurred in favor of the non-material, spiritual sacrifice (sacrificing personal time and relations etc.).
13 A similar approach happens when pro-life/pro-vita organizations discuss the health risks which may occur in the case of women who have abortions. It is true that any abortion has risks, but what is being omitted is the fact that any pregnancy has risks. Therefore, the pregnant woman is prone to greater risks than the non-pregnant woman.

References

Ariès, Philippe. 1965. *Centuries of Childhood: A Social History of Family Life*, trans. R. Baldick. New York: Random House.

Badinter, Elizabeth. 2012. *The Conflict. How Modern Motherhood Undermines the Status of Women*, trans. A. Hunter. New York: Metropolitan Books.

Baumeister, Roy. F. 1991. *Meanings of Life*. New York: Guilford Press.

Berger, Peter. L., and Thomas Luckmann. 1966. *The Social Construction of Reality: A Treatise in the Sociology of Knowledge*. New York: Doubleday.

Bologna, Caroline. 2014. "22 Candid Photos That Show How Beautiful Breastfeeding Really Is". *The Huffington Post*, June, 8. www.huffingtonpost.com/2014/08/06/public-breastfeeding-awareness-project-photos_n_5651252.html (accessed August 10, 2015).

Brown, Ivana. 2010. "Ambivalence, Maternal." In *Encyclopedia of Motherhood*, ed. Andrea O'Reilly. Thousand Oaks, CA: Sage.

Brown, Petrina. 2004. *Eve. Sex, Childbirth and Motherhood through the Ages*. West Sussex: Summersdale.

Carr, Glynis. 2011. Forward to *Ecofeminism and Rhetoric. Critical Perspectives on Sex, Technology, and Discourse*, ed. Douglas A. Vakoch. New York and Oxford: Berghahn.

Crittenden, Ann. 2001. *The Price of Motherhood: Why the Most Important Job in the World Is Still the Least Valued*. New York: Henry Holt.

Cronan Rose, Ellen. 1991. "The Good Mother: From Gaia to Gilead, Frontiers," *A Journal of Women Studies* 12(1): 77–97.

Cuomo, Chris. J. 1998. *Feminism and Ecological Communities: An Ethic of Flourishing*. New York: Routledge.

Curtis, Polly. 2014. "Caesarean or Natural Birth—Which is Safer?" *The Guardian*, May 20. www.theguardian.com/politics/reality-check-with-polly-curtis/2011/nov/23/health (accessed August 15, 2015).

d'Eaubonne, Françoise. 1974. *Le Féminisme ou La Mort*. Paris: Pierre Horay.

Dainius. 2015. "Stunning Photos of Moms Breastfeeding Outside Show Nursing in Public Is OK." *Boredpanda,* June. www.boredpanda.com/public-breastfeeding-family-photography-goddess-ivette-ivens/ (accessed August 11, 2015).

Dawkins, Richard. 1995. *River Out of Eden*. New York: Basic Books.

De Angelis, Massimo. 2007. "Production and Reproduction." In *The Beginning of History. Value Struggles and Global Capital*, ed. M. de Angelis. London and Ann Arbor: Pluto Press.

De Souza, Ronald. 2004. "Enfanter et mourir. Réflexions sur une approche biologique du rapport entre la sexualité et la mort." In *La mort et l'immortalité. Encyclopédie des savoirs et des croyances*, ed. F. Lenoir and J.-P. De Tonnac. Paris: Bayard.

Douglas, Susan. J., and Meredith W. Michaels. 2004. *The Mommy Myth: The Idealization of Motherhood and How It Has Undermined Women*. New York: Free Press.

Dunnewold, Ann L. 2007. *Even June Cleaver Would Forget the Juice Box: Cut Yourself Some Slack (and Still Raise Great Kids) in the Age of Extreme Parenting*. Arlington: Health Communications.

Durand, Gilbert. 1960. *Les structures anthropologiques de l'imaginaire*. Paris: Presses Universitaires de France.

Elias, Norbert. 2001. *The Loneliness of the Dying*. London and New York: Continuum International.

Erin White Photography. 2015. www.erinwhitephotography.com (last modified August 2015).

Evenson, Ranae. J., and Robin. W. Simon. 2005. "Clarifying the Relationship between Parenthood and Depression," *Journal of Health and Social Behavior* 46(4): 341–358.

Faludi, Susan. 2006. *Backlash: The Undeclared War against American Women*. New York: Three Rivers Press.

Feinmann, Jane. 2011. "Is an Obsession with Natural Birth putting Mothers and Babies in Danger?" *Daily Mail*, October, 4. www.dailymail.co.uk/health/article-2044875/Is-obsession-natural-birth-putting-mothers-babies-danger.html (accessed August 15, 2015).

Glaser, Eliane. 2015. "The Cult of Natural Childbirth Has Gone Too Far." *The Guardian* March 5. www.theguardian.com/commentisfree/2015/mar/05/natural-childbirth-report-midwife-musketeers-morcambe-bay

Gotlib, Anna. 2010. "'Bad' Mothers." *In Encyclopedia of Motherhood*, ed. A. O'Reilly. Los Angeles: Sage.

Growingupawife. 2013. "I'm Sick of Apologizing for My C-section." *BlogHer*, April 24. www.blogher.com/c-section-awareness-month-my-perspective (accessed August 15, 2015).

Hakim, Catherine. 2000. *Work-Lifestyle Choices in the 21st Century*. Oxford: Oxford University Press.

Hansen, Michèle, Carol Bower, Elizabeth Milne, Nicholas de Klerk, and Jennifer J. Kurinczuk. 2013. "Assisted Reproductive Technologies and the Risk of Birth Defects – a Systematic Review." *Human Reproduction* 20(2): 328–38.

Hays, Sharon. 1998. *The Cultural Contradiction of Motherhood*. New Haven: Yale University Press.

Hernandez, Marcia. 2010. "Capitalism and Motherhood." In *Encyclopedia of Motherhood*, ed. Andrea O'Reilly. Thousand Oaks, CA: Sage.

Hrdy, Sarah Blaffer. 1999. *Mother Nature: A History of Mothers, Infants, and Natural Selection*. New York: Pantheon Books.

Ifeanyi, K. C. 2015. "You'll Never Look at Public Breastfeeding the Same After You See This Gorgeous Photo Series." *Fast Company*, June 18. www.fastcompany.com/3047606/the-recommender/youll-never-look-at-public-breastfeeding-the-same-after-you-see-this-gorgeou (accessed August 10, 2015).

International Cesarean Awareness Network. 2015. "What Does the Burgundy Ribbon Stand For?" www.ican-online.org/blog/sp_faq/what-does-the-burgundy-ribbon-stand-for/ (last modified August 2015).

Ivens, Ivette. 2015. www.ivetteivens.com (last modified August 2015).

Jaclyn. 2014. "Cesarean Sections: Leaving More Than Just a Physical Scar (One Woman's Journey to a Better Way to Birth)." *Natural Momma*, April 29. www.naturalmommainprogress.com/2014/04/29/cesarean-sections-leaving-just-physical-scar-one-womans-journey-better-way-birth/ (accessed August 15, 2015).

Jong, Erica. 2010. "Mother Madness." *The Wall Street Journal*, November, 6.

Klinenberg, Eric. 2013. *Going Solo. The Extraordinary Arise and Surprising Appeal of Living Alone*. New York: Penguin Press.

Lad-Taylor, Molly, and Lauri Umansky, eds. 1998. *"Bad" Mothers: The Politics of Blame in the 20th-Century America*. New York: New York University Press.

Lamaze International. 2015. "Vaginal Birth After Cesarean (VBAC): Informed and Ready". https://elearn.lamaze.org/courses/vbac-informed-and-ready (last modified August 13, 2015).

Lamaze, Fernand. 1984. *Painless Childbirth: The Lamaze Method*. Chicago: Contemporary Books.

Le Breton, David. 1990. *Anthropologie du corps et modernité*. Paris: Presses Universitaires de France.

Leboyer, Frederick. 2009. *Birth without Violence*. Rochester: Healing Arts Press.

Leyser, Ophra. 2010. "Infertility". In *Encyclopedia of Motherhood*, ed. A. O'Reilly. Los Angeles: Sage.

Liao, Marina. 2015. "38 Timeless Photos of Moms Breastfeeding Their Children at Every Stage." *Popsugar*, August 8. www.popsugar.com/moms/Photos-Breastfeeding-36178608#photo-37580230 (accessed August 10, 2015).

Liedloff, Jean. 1986. *The Continuum Concept: In Search of Happiness Lost*. New York: Perseus Books.

Lifton, Robert. J., and Eric Olson. 1974. *Living and Dying*. London: Wildwood House.
McLanahan, Sara S., and Erin L. Kelly. 2006. "The Feminization of Poverty: Past and Future." In *Handbook of the Sociology of Gender*, ed. J. S. Chafetz. New York: Springer.
Muza, Sharon. 2015. "World Health Organization: Provide Cesareans for Women in Need, Don't Focus on Specific Rate." *Science & Sensibility. A Research Blog About Healthy Pregnancy, Birth & Beyond from Lamaze International*, April 21. www.scienceandsensibility.org/cesarean_awareness_who-report/ (accessed July 20, 2015).
Odent, Michel. 1994. *Birth Reborn: What Childbirth Should Be*. London: Souvenir Press.
Purcel, Fiona. 2015. "18 Stunning Photos of Breastfeeding Mothers." *Mother & Baby*, May 18. www.motherandbaby.com.au/baby/feeding/2015/5/18-stunning-photos-of-breastfeeding-mothers/ (accessed August 11, 2015).
Rochman, Bonnie. 2011. "Working Women Who Try to Be 'Supermom' May Be More Depressed." *Time* August 23. http://healthland.time.com/2011/08/23/working-women-who-subscribe-to-the-supermom-myth-are-more-likely-to-be-depressed/
Shiva, Vandana. 2005. *Earth Democracy*. Cambridge, MA: South End Press.
Sjoo, Monica and Barbara Mor. 1987. *The Great Cosmic Mother: Rediscovering the Religion of the Earth*. San Francisco: Harper.
Sowards. Stacey K. 2011. "Gender Representations in Orangutan Primatological Narratives: Essentialist Interpretations of Sexuality, Motherhood, and Women." In *Ecofeminism and Rhetoric: Critical Perspectives on Sex, Technology, and Discourse*, ed. D. A. Vakoch. New York and Oxford: Berghahn Books.
Teodorescu, Adriana. 2012. "Symbolic Immortality through Nature. The Deconstruction of a Myth: A Thanatological Perspective," *Philobiblon. Transylvanian Journal of Multidisciplinary Research in Humanities* 17(2): 378–402.
Terreri, Cara. 2015. "Because." *Science & Sensibility. A Research Blog About Healthy Pregnancy, Birth & Beyond from Lamaze International*, April 9. www.scienceandsensibility.org/cesarean-awareness-month/ (accessed July 10, 2015).
Thurer, Shari. 1995. *Myths of Motherhood: How Culture Re-invents the Good Mother*. New York: Penguin.
Tuteur, Amy. 2011. "Ten Lies My Natural Childbirth Educator Told Me." *The Skeptical OB*, February 23. www.skepticalob.com/2011/02/ten-lies-my-natural-childbirth-educator.html (accessed August 15, 2015).
Van Tongeren, Paul. 1995. "The Paradox of Our Desire for Children." *Ethical Perspectives* 2: 55–62.
Vireday, Pam. 2014. "Practice Variation in Cesarean Rates: Not Due to Maternal Complications." *Science & Sensibility. A Research Blog About Healthy Pregnancy, Birth & Beyond from Lamaze International*, November 13. www.scienceandsensibility.org/page/2/?s=cesarean (accessed July 20, 2015).
Warner, Judith. 2006. *Perfect Madness. Motherhood in the Age of Anxiety*. New York: Riverhead Trade.
Warren, Karen J. 1987. "Feminism and Ecology: Making Connections," *Environmental Ethics* 9(Spring): 3–21.
World Health Organization. 2015. "Caesarean Sections Should Only Be Performed When Medically Necessary." www.who.int/mediacentre/news/releases/2015/caesarean-sections/en (last modified August 2015).
World Health Organization. 2015. "Maternal Mortality." www.who.int/mediacentre/factsheets/fs348/en (last modified August 2015).
Zurn, Krista. 2015. "The Shame Game." *Erin White Photography* 15 May. www.erinwhitephotography.com/the-shame-game (accessed August 10, 2015).

6 The nature of body image: the relationship between women's body image and physical activity in natural environments

Denise Mitten and Chiara D'Amore

Body image is a significant personal and societal issue, especially for women. Numerous studies have shown that the substantial majority of adult women have a negative relationship with their bodies, an area of concern that starts early in childhood for many girls. Not only are there detrimental personal emotional, psychological, and physical effects of such pervasive negative body image, but there are also numerous societal and even ecological implications of this epidemic. People who are indoctrinated by cultural norms to feel like their bodies are inferior to beauty ideals invest tremendous mental and emotional energy as well as time and economic resources toward trying to achieve often unattainable expectations. Therefore the achievement of a positive body image frees up those resources to be put toward greater personal and societal good. As powerfully shared by Sally McGraw (2012):

> Learning to love your body may seem small or selfish or pointless at times, especially when compared to fighting for larger causes and reaching out to help others. But to fashion yourself into a powerful, effective, whole being, you've got to come at life from a place of strength. Your body is your home. If you hate your home – if you flee from it, disrespect it, and wish it were fundamentally different – your strength will be diminished. Whether you want to help others or simply find your way to happiness in your own life, loving yourself is absolutely vital. And loving yourself includes loving your body. Your body is integral to yourself.

This chapter begins with an overview of body image—what it is, how it has evolved, and why it is a significant personal and societal issue. It proceeds to briefly describe the relationship between humans and the natural environment, the effects of lifestyles that are increasingly 'disconnected' from nature, and the increasingly well-understood importance of direct experiences in nature for human well-being. Bringing these two topics together, the discussion hones in on how being active outdoors has a positive effect on women's body image. The results of recent and ongoing research conducted by the authors on the effects of being physically active in natural environments on women's perceptions of their physical effectiveness and attractiveness are presented as offering important insight into this topic. For example, our research has found that women who spent more than three hours a week engaged in nature-based activities had a significantly more positive body image than their counterparts. The discussion uses ecofeminism scholarship to suggest that patriarchal beliefs contribute to the diminishing and objectifying of women and nature and that both women and nature would benefit from an ecofeminist paradigm. Through time in nature some of the negative impacts of patriarchal beliefs about women can be healed and women gain new perspectives. Time in nature, including time in women's groups, helps women achieve the characteristics that other research has determined to contribute to sustaining positive body image. By healing our internal relationship through time in nature, we are better able to help heal the human–nature relationship.

Women's body image

Body image refers to the way people perceive their own body: their perceptions about how other people view their body, and the complex relationship between these internal and external perceptions (Robinson 2003). Body image is an aspect of self-concept—the global and multidimensional perception of who one is—including one's actual experiences and the interpretations about those experiences (Kaiser 1997). The construct of body image contains both positive and negative elements and includes features related to perceptions (cultural standards of beauty and the way in which a person does or does not match these standards), attitudes (the relative personal and societal importance placed on that match), and behaviors (efforts made to adhere to standards of beauty) (Fallon 1990). Body image can be a basic litmus test of self-affirmation, self-worth, self-efficacy, and self-love.

In most recent cultures, women have been told to reshape their body or to wear garments, such as corsets, that hide or bind some parts of the body while accentuating other body parts. In contemporary history these unrealistic standards for women's bodies have intensified—both in physical expectations and in the saturation of messages around what constitutes beauty. Until the early 1900s, in Western cultures it was often considered to be a sign of health and wealth for a woman to have a Rubenesque body type. Around this time consumer, culture began to increasingly influence standards of beauty for women's bodies through cosmetics, fashion, cinema, and advertisements. As a result, the preference in the United States shifted from a more voluptuous female form to a thinner frame with fewer curves. Exercise began to be viewed as an activity necessary to achieve the level of thinness that had become the new sign of wealth (Cash and Pruzinsky 2004). The global reach of Western media has a pernicious effect on women's body image around the world. A Harvard Medical Center study in Fiji found that, in a culture that long considered robust girls and women healthier and stronger, three years after the introduction of TV shows such as *Beverly Hills, 90210*, 74 percent of Fijian teenagers described themselves as fat, with those who watched TV three or more time a week being 30 percent more likely to go on a diet (Becker 2004). With this focus on thinness came an increase in the practice of dieting for the purpose of losing weight.

In the 1960s the waif-like look became popularized by the supermodel Twiggy Lawson, who was arguably the first underweight woman to become the standard for the ideal body. It was also at this time that Barbies, with their impossible body proportions, became a popular toy for young girls. Since the 1980s, and continuing through today, the ideal female body type popularized in the media is ultra-thin and toned. To illustrate how unattainable the current thinness ideal is for most women, consider that the average U.S. female model is 5'11" and weighs 117 pounds while the average woman in the U.S. is 5'4" and 140 pounds—only two percent of women in the U.S. are model sized (Katz 2004). Thinness has come to represent beauty and a higher socio-economic status as well as to symbolize success and self-control (Forehand 2001).

Many women have internalized society's thinness ideal, and when they cannot achieve the standards set for the female body, they often develop a negative body image. A number of authors (Coker and Abraham 2014; Matthiasdottir, Jonsson, and Kristjansson 2012; Peat, Peyerl, and Muehlenkamp 2008; Sinclair 2006; Tiggemann 2004) have reported that body dissatisfaction is present in the majority of women in developed countries, including women at normal weights. Body dissatisfaction is defined as the negative and dysfunctional beliefs and feelings about one's shape and weight (Garner 2002). Such poor body image starts at disturbingly young ages, with 51 percent of 9 and 10 year old girls in the U.S. reporting that being on a diet makes them feel better about themselves (Spake 2004). By the time they reach adulthood, approximately 80 percent of women in the U.S. are not satisfied with their bodies (Katz 2004).

Over thirty years ago the term *normative discontent* was coined to describe the Western cultural norm for women who have negative views of their bodies (Rodin, Silberstein, and Striegel-Moore 1984). Negative body image has far reaching consequences—impacting many aspects of women's self-concept and wellness, including self-esteem, self-efficacy, personal growth, positive relations, and purpose in life; all of which have implications for personal, family, community, and environmental well-being (Cash 1998). For example, female adolescents in Poland who were extremely critical of their bodies were found to have fewer life skills because of a lower trust in their possibilities, a desire to be someone else and change their appearance, and less frequently perceived happiness (Laudanska and Bronikowski 2013). This dissatisfaction seems consistent for all ages of women, though the negative effects for some women on their self-concept lessens with age (Webster and Tiggemann 2003). A negative body image can lead to or exasperate a variety of mental health problems, including depression, anxiety, body dysmorphic disorder, and eating disorders (Buddeberg-Fischer, Klaghofer, and Reed 1999; Coker and Abram 2014). For example, when comparing adolescents with psychiatric illnesses, it was found that the individuals with negative body image tended to be more depressed, anxious, and suicidal than those without the same dissatisfaction about their appearance (Dyl et al. 2006).

Body dysmorphic disorder and eating disorders, such as anorexia and bulimia nervosa, are conditions defined by abnormal eating habits that may involve either insufficient or excessive food intake to the detriment of the individual's physical and mental health. These chronic mental illnesses can be life threatening where the afflicted individual has excessive concern about and preoccupation with a perceived defect of their physical appearance. Body dysmorphic disorder often co-occurs with depression, anxiety, social withdrawal, and isolation. Negative body image and the often-accompanying mental concerns affect physical, social, and emotional domains in life and may cause problems in home, social, and work environments.

Theories about how females are influenced to have internalized body dissatisfaction and strive for the thin-ideal include social comparison theory (Festinger 1954) and objectified body consciousness theory (Fredrickson and Roberts 1997; McKinley and Hyde 1996). Social comparison theory says that humans strive to assess their opinions, progress, and standing in life through comparison with others. Wheeler and Miyake (1992) found that social comparisons on the dimension of physical appearance tend to be upward rather than downward and that comparisons often result in a decrease in self-perceptions of attractiveness. Morrison, Kalin, and Morrison (2004) found that social comparison is a powerful predictor of body image evaluation for female adolescents. Objectified body consciousness claims that the "feminine body is socially constructed as an object to be looked at" and contends that when sexual objectification is experienced, women become socialized to engage in this kind of objectification toward themselves and others (Sinclair 2006, 50). This theory argues that as a woman internalizes the social 'demands' on her body, she begins to see her own physical body as an object that is measured, evaluated, and valued through the lens of another person's view of her. Sinclair (2006, 51) explained the results of objectified body consciousness as follows: constant monitoring of how one's body looks (body surveillance); internalization of cultural body standards (body shame); the belief that one's appearance can be controlled (appearance control beliefs). Another example of objectified body consciousness is what Maine and Kelly (2005) refer to as the body myth. Body myth describes the widely held belief that one's self-worth and worth to others is, and should be, determined by how we look, how much we weigh, and the food we eat.

Thus far we have articulated precipitating factors and problems related to a negative body image. Now using a reparative approach to relate to nature, we turn to how one might help create a positive body image. More than merely mitigating the effects of negative messages about one's body image, time in nature may promote a positive body image and may provide inoculation against media messages of the thin-ideal. Tylka and Wood-Barcalow (2015) commented that, if we focus on eliminating negative body image without understanding how to promote positive body image, then clinicians and others may be less able to help women attain health in this area. They said, "Helping clients adopt a positive body image may help them appreciate, respect, celebrate, and honor their bodies, which may make treatment gains more effective and lasting" (Tylka and Wood-Barcalow 2015, 118).

Recent studies focusing on characteristics and causes for positive body image found that women who enjoy physical activity, have increased social support, appreciate their body functioning, have unconditional acceptance from others, and broadly conceptualize beauty are likely to experience positive body image (Avalos, Tylka, and Wood-Barcalow 2005; Williams, Cash, and Santos 2004; Wood-Barcalow, Tylka, and Augustus-Horvath 2010). Many women experience these sources of positive body image when in the outdoors.

The human–nature relationship

People have spent over 99 percent of their evolutionary history in close proximity to and in relationship with nature. This intimate relationship with nature was based on what Flinders (2003) described as a *values-of-belonging*, which means that a society's core values include intimate connection to and empathetic relationship with all beings, generosity, egalitarianism, and non-violent conflict resolution. Although some cultures never entered an agrarian lifestyle and thus remain hunter-gathers (such as the Bushmen of the Kalahari and the Betek people living in forests of Peninsular Malaysia), and although some have remained predominately peaceful groups, overall, the last few hundred years have wrought an extraordinary disengagement of humans from the natural environment. This shift is especially pronounced for people in Westernized and/or industrialized countries—those that have the greatest historical culpability for environmental issues. Significant historical shifts in human culture, such as the domestication of plants and animals, monotheism, the 18th-century European Enlightenment, and the industrial revolution, played particularly significant roles in the strengthening of a patriarchal ideology that created a duality in which "women are treated as inferior to men, 'nature' is treated as inferior to 'culture', and humans are understood as being separate from, and often superior to, the natural environment" (Harris 2015, 1). The result has been tremendously damaging for the well-being of both the natural environment and women.

Today people in the U.S. spend more than 90 percent of their time indoors, a level of constraint rare even in recent generations (Clements 2004). As our lived experiences in direct contact with nature rapidly wane, evidence is mounting of the critical connections between human well-being and time spent in a healthy natural world. Contact with nature is essential for our physical health, as well as for our optimal cognitive, psychological, and emotional well-being (Ewert, Mitten, and Overholt 2014). Research has examined the impact of exposure to nature in a variety of forms, from mere images, to views from a window, to neighborhood play, to educational settings and wilderness programs. The populations studied have been from a wide range of demographics, socio-economic groups, and countries, and have included the general public as well as clinical populations. The specific outcomes examined have been diverse as well, including physical, cognitive, emotional, and social development. The outcome of positive findings from studies employing different populations, research designs, and measures suggests that the benefits of time in nature are pervasive and generalizable. The following is a brief summary of the evidence on the positive effect of contact with nature for human health and well-being.

Physical benefits

The physical benefits of time in nature are increasingly well documented and frequently linked with the increase in physical activity that comes with time spent outdoors (Cleland et al. 2008; Potwarka et al. 2008) as well as uneven trails that help increase balance. For example, Lee and colleagues (2014) found lower heart rate and systolic blood pressure levels for people walking in the forest verses people walking in urban environments. Li and colleagues (2008) found that a forest bathing trip (walking in the forest) can increase human natural killer NK activity, the number of NK cells in female subjects, which is seen as aiding the immune system. Conversely, a decrease in time spent outdoors has been found to contribute to physical ailments such as obesity, hypertension, and myopia.

Cognitive benefits

Cognitive benefits from time in nature include creativity, problem-solving, focus, and self-discipline. Walking in nature for fifteen minutes (in comparison to walking in an urban environment) has been found to increase an individual's subjective connectedness to nature, positive affect, attentional capacity, and ability to reflect on a life problem (Mayer et al. 2009). In five studies, Ryan and colleagues (2010) found that nature exposure relates to both physical and mental vitality.

Psychological benefits

Nature relatedness has been found to be significantly correlated with the six dimensions of psychological well-being—autonomy, environmental mastery, positive relations with others, self-acceptance, purpose in life, and personal growth for children and adults (Nisbet, Zelenski, and Murphy 2011). As noted by Weinstein, Przybylski, and Ryan (2009), nature can bolster autonomy directly by affording opportunities for introspection and a coherent sense of self (Walker, Hull, and Roggenbuck 1998) and providing an alternative to the pressuring elements of everyday life (Stein and Lee 1995). Nature connectedness is associated with mindfulness (Howell et al. 2011), which is in turn supportive of self-awareness, self-esteem, and resilience (Coholic 2011) and reduces maladaptive rumination (Heeren and Philippot 2011). An effective connection with nature has been found to foster an overall sense of psychological well-being (Korpela et al. 2009).

Emotional benefits

Emotional benefits include stress reduction, reduced aggression, and increased happiness (Chawla 2006). Kuo and Sullivan (2001) found that levels of aggression and violence, including domestic abuse, were significantly lower among individuals who lived near some natural elements than among their counterparts who lived in barren conditions. Chiesura (2004) found that people visiting an urban park perceived regeneration of emotional equilibrium, relaxation, and the stimulation of a spiritual connection with the natural world as key benefits from their experience. Hug and colleagues (2008) compared the restorative effects of physical activities in forest and indoor settings. While the respondents viewed physical exercise in either location as beneficial, those exercising in a forest landscape reported a greater improvement in their mental balance and release from everyday hassles.

Taken together, the aforementioned studies demonstrate the positive relationship between time spent in nature, feeling connected to nature, and human health and well-being, all of which have positive implications for the ability to heal or inoculate against negative body image. The following section presents research about the relationship between outdoor activity and women's body image.

Women's body image and outdoor activity

For twenty years Dr. Mitten ran Woodswomen, Inc., an organization that offered outdoor trips for women and children. In this role she witnessed the journeys of more than 7,000 women as they experienced one day to four weeks in the outdoors. She heard many stories from women who felt healed or experienced a sense of coming home when traveling and living in the natural environment and who had positive, life-changing experiences, including increases in self-esteem, empowerment, personal control, self-efficacy, physical strength, and self-care (Mitten 1992).

In 1997, West-Smith (2000) began surveying women who had gone on outdoor adventures with Woodswomen to study the effect of participating on outdoor trips on body image. Her study demonstrated that women who participated in outdoor adventures at least monthly for a year had more positive body image compared with those who were not active in outdoor adventures. Through a combination of quantitative and qualitative analysis, West-Smith discovered that most active outdoorswomen are satisfied with their physical attractiveness and their bodies' physical effectiveness to a greater degree than the general U.S. population.

Woodruff (2009) surveyed women over forty years old who participated in a five-day outdoor trip in a different program and found that they had an increased positive body image as compared with a control group of women not engaging in activities in the outdoors. Specifically, as their level of outdoor activity increased, the more important it was for these study participants to be physically effective. The participants defined physical effectiveness as having intrinsic motivation and physical capability. Participants indicated that, for them, attractiveness and effectiveness were interconnected, and being physically attractive meant that one possessed a combination of physical ability, inner confidence, expressed self-assuredness, and physical proportion. This broader definition counteracts the societal norm of preferring an ultra-thin body and objectifying the physical body. These results echo other research demonstrating that outdoor experiences can help women create a positive body/self-relationship in part because active outdoorswomen are able to rebuff cultural and stereotypical definitions of beauty and, as a result, maintain a more positive body image (Hornibrook et al. 1997; Kiewa 1996, 2000; McDermott 2004; Mitten 1992; Pohl, Borrie, and Patterson 2000; Whittington 2006; Mitten and Woodruff 2010).

There may be an added impact when women are in the natural environment with other women. Both the West-Smith (1997) and Woodruff (2009) studies reported that women perceived benefits from being in women's groups. Fredrickson and Anderson (1999) found in their study of two women's groups that both the natural area and being in an all-women's group were important for women's positive feelings:

> Many testified to having grown both physically and emotionally from having had the opportunity to share their life's joys and sorrows with other women who were willing to "really listen" and to be nonjudgmental, and who were also actively supportive of what they had experienced both physically and emotionally while on the trip.
>
> (29)

Additionally, they specifically mentioned the non-competitive atmosphere, group trust and emotional support, sharing common life changes, the opportunity for direct contact with nature, periods of solitude, and the inherent physical challenges. They reported that the wilderness setting had a major influence on more contemplative aspects of the participants' experiences and that the social dynamics between and among group members contributed to the spiritually beneficial aspects of the experience. Furthermore the strong group cohesion and emotional safety between participates were paramount to most participants' experiences which in turn influenced the spirituality inspirational qualities of the place setting.

The aim of the study recently conducted by the chapter authors was to build on existing research and develop a more robust understanding of the relationship of women's body image and their time spent being active in the natural environment. This study brought to this area of inquiry a larger and more varied population of women than preceding studies and honed in on various aspects of outdoor activity, such as time, frequency, and duration in order to relate these factors to body image. The study utilized a self-report questionnaire developed from an instrument created by West-Smith (1997) and modified by Woodruff (2009). The core of the questionnaire was the Body Cathexis Scale (BCS), developed by Secord and Jourard in 1953 and still a widely used tool to assess the degree of satisfaction or dissatisfaction with various parts and features of the body (Grogan 1999). Subjects evaluate forty-six body parts and functions according to a five-point Likert scale ranging from 1, 'strong negative', to 5, 'strong positive', resulting in a total score that indicates overall body satisfaction. The BCS has been established in reports of a test-retest coefficient as having a reliability of between .84 and .91 and .96 for the current study.

Using the online survey tool, Survey Monkey, the forty-four-item questionnaire used in this study was divided into four parts: Section 1 asked for outdoor activity information; Section 2 sought body data including the BCS and questions about physical attractiveness and physical effectiveness; Section 3 focused on women's life experiences, such as relationship status, child-bearing, and history of eating disorder/cosmetic surgery; and Section 4 gathered demographics. The questionnaire was distributed during the spring of 2013 to the 625 women that were enrolled at Prescott College, a small liberal arts college in Arizona that has an emphasis on adventure education and field-based courses. A total of 186 women completed the survey sufficiently for analysis, which was a 30 percent response rate. The women in this study ranged in age from nineteen to sixty-five years old, with an average age of thirty-four, and were studying in undergraduate or graduate programs. Means were computed for the BCS followed by a t-test for equality of means for a variety of variables such as age, education, quantity and duration of outdoor activity, importance of being physically attractive and effective, and self-rated level of physical attractiveness and effectiveness.

BCS mean

The women that participated in this study had a higher perceived body image (BCS mean=3.45) than the expected average U.S. population of women (2.5) and of the control group in Woodruff's study (3.31). This study population's BCS mean results were not statistically different than in West-Smith's group of active outdoor women's (3.42) or Woodruff's group of short-term adventure program participants (3.51) as illustrated in Table 6.1.

Table 6.1 A comparison of BCS mean results for three studies about women's body image

Study Population	*BCS Mean*
Woodrufff's (2009) Control group (n=17)	3.31
West-Smith's (1997) group (n=86)	3.42
Females at Prescott College (n=186)	3.45
Woodrufff's (2009) Short term adventure group (n=39)	3.51

No statistically significant correlation between the respondent's age and their BCS mean was found, which is congruent with other research results (Webster and Tiggemann 2003). Analysis of the respondent's education level also found no statistically significant correlation with their BCS mean.

Quantity of outdoor activity

When asked, "About how much time do you spend in outdoor activity in a current typical week?" the responses ranged from zero to fifty hours per week with an average of ten hours per week. Table 6.2 illustrates the thresholds of time spent active in the outdoors that had a statistical significance on the BCS for this study population, demonstrating a correlational trend that as the amount of time spent participating in outdoor activities increases so does a woman's positive perception of her body image.

Duration of time outdoors

In this study, 96 percent of the participants had spent at least one overnight outdoors and 63 percent of respondents had spent more than two weeks outdoors one or more times. For participants in this study, the impact of time spent outdoors on body image was a positive trend in the BCS mean as people experience greater duration of concurrent overnights outdoors, number of overnight trips, and total number of overnights outdoors. Further, the average BCS was lowest for people who have never experienced an overnight outdoors and increased with the experience of one to three nights outdoors and with the experience of four or more concurrent nights outdoors. Similarly, the BCS average trend was higher the greater the number of overnight trips reported as well as the total number of nights the women reported spending outdoors.

Physical attractiveness and effectiveness

For participants in this study, there was a strong positive correlation between perceptions of personal physical effectiveness and their BCS mean (i.e., *r* of .00097 for effective vs. all ineffective categories). This indicates that as the women's level of participation in outdoor activity increased, the more important it was for them to be physically effective. This correlation remained whether or not the woman believed that physical attractiveness or physical effectiveness

Table 6.2 Thresholds of significance for time spent in outdoor activity for a population of women students in higher education

< 3 hours (3.26 BCS)	vs.	> 3 hours (3.5 BCS)	= *r*(0.024)
< 4 hours (3.29 BCS)	vs.	4+ hours (3.5 BCS)	= *r*(0.037)
< 8 hours (3.36 BCS)	vs.	8+ hours (3.55 BCS)	= *r*(0.023)
≤ 10 hours (3.38 BCS)	vs.	> 10 hours (3.63 BCS)	= *r*(0.010)
< 15 hours (3.38 BCS)	vs.	15+ hours (3.67 BCS)	= *r*(0.008)

were important. There also was the expected significant correlation between a woman's perception of her physical attractiveness and her body image. In other words, the women who believe they are physically attractive have a higher body image as evidenced by their BCS mean. The following are select quotes from study participants:

> When I am able to do the things I want to do I feel more confident. I think that people who exude confidence are more often perceived as attractive regardless of their physical attributes.

> When I feel healthier and more effective, I feel more positive and engage in life more freely. This impacts my relationships and connections to others. I feel stronger and more attractive therefore more willing to be out in the world.

> I used to have an eating disorder and considered myself hideous. This prevented me from being active because I couldn't see my body as anything but useless and ugly. I believed working out was a waste of my time because I didn't think my body could be anything other than what I perceived it as. After months of counseling, I took a trip to Peru. While there, I did a 30 kilometer hike over two and a half days and hiked above 8,000 ft in altitude. When I returned to Cuzco after the trek, I viewed my body in a whole different light; while I hadn't lost significant weight, I suddenly viewed my "fat" thighs as strong and supportive, my tool that carried me through that experience. Being outside and pushing my body physically changed how I viewed it mentally.

The results of this study support that through outdoor activity, self-concept may increase, beliefs about attractiveness change, and psychosocial variables about the ideal body may be mitigated. This means that the external influence of the socially or culturally preferred body type had less influence on this population of women than their internalized desire for engaging in outdoor activities and their self-determined need for physical effectiveness to do so. In previous studies, the same relationship was found (Mitten and Woodruff 2010; West-Smith 1997). Specifically, as women's level of participation in outdoor activities increased, the more important it was for them to be physically effective and the less they worried about culturally prescribed physical attractiveness.

An ecofeminist discussion

Throughout history, the Earth has been described in the feminine. Both physiological (i.e., menstrual cycles linked with lunar cycles, the act of creation that is pregnancy and childbirth) and psychological (i.e., nurturant, cooperative) attributes of women are linked with the ways of nature and in many ancient and some current cultures these attributes are revered. However, with the advent of patriarchal beliefs, relationships have fundamentally changed from an ethic of valuing all people to valuing some people, behaviors, and resources more than others.

Patriarchal thinking, rooted in a dualistic world view, splits mind from body, spirit from matter, male from female, culture from nature. One concept in each pair is deemed superior to the other. This paradigm has caused people in industrialized parts of the world to steadily become more physically, mentally, and emotionally estranged from the nature, perceiving it as a commodity solely for human use; there has been a concurrent continuation in the diminishment of the inherent worth of women and the feminine. In such societies the commodification and objectification of nature and of women are similar and come from giving entitlement to what is labeled or considered masculine, which leads to domination and power and control over others.

Sexism as well as many aspects of racism, classism, imperialism, speciesism and other forms of prejudice operates through this hierarchy. Baran (Chapter 3, this volume), using Collins's (1993) work, describes the effects of the intersectionality of these prejudices in more detail. Causes of negative body image can be traced to dualistic thinking and prejudice, which allow for objectification and the devaluing of what is labeled feminine. In contrast, ecofeminism fosters a sense of our belonging to, rather than being in control of, the community of life. As a scientist, pragmatic feminist, and early ecofeminist, Ellen Swallow Richards intended to accentuate interconnectedness when, in 1882, she introduced the name "ecology" to the English language (Clarke 1973). She understood the connection between the environment and human health and that we needed to change the way we related to the environment in order to maintain our health as well as the health of the environment of which we are a part. She recognized that we are entangled with nature and that human quality of life is dependent on society's capacity for teaching its members how to coexist harmoniously with their environment, including their family, community, and world. However, ecology as a discipline became less about humans' relationship to our environment and more about a mechanical interpretation of nature.

Françoise d'Eaubonne (1974), a French activist, who coined the term 'ecofeminism,' recognized, as Swallow Richards did, that by externalizing and treating the environment in ways that were violent, uncaring, and destructive we actually hurt ourselves. She understood how impacts on the biosphere and women's reproductive rights were concrete manifestations of the intersections of feminism and ecology. She and others urged people to see the link and to change the patterns of violence, which means changing our relationships. Eco, from the Greek word *oikos*, denotes the whole household of life; the term 'ecofeminism' therefore means to care for all life. Valera (Chapter 1, this volume) said, "thanks to d'Eaubonne, we can say today that the expropriation of the female body and Nature fall under a single dominant approach, which has led to almost irreparable damages." Valera concurs with Swallow Richards and d'Eaubonne, stating, "Ecofeminism once again tends to combine elements of feminism (the adversity to dichotomies) and ecologism (the holistic vision of all), creating a more complete picture of reality."

Ecofeminism and ecology reflect how our relationships are intertwined with all beings and natural systems as an ecology of relationships (Mitten 2016). Relationships are central to every dimension of our lives and shape our lives because we are social, relational beings and because feedback loops continually operate in relationships. Relationships govern our lives because we are entangled with the whole cosmos from the microbes that live in our guts to the sun on whose light we depend. Our understanding of relationships and skill in developing and maintaining healthy relationships is central to maintaining personal and societal health and well-being. This includes relationships with others as well as the natural world, and a healthy relationship with self.

Being in nature offers women new or different perspectives and opportunities to explore these perspectives in relationship to self, others, and the natural environment. It offers women different information than they have had access to and different ways of knowing. These experiences can lead to transformations in relationships with self, others, and the more-than-human world. Re-discovering how to be in healthy relationships, including being in relationship with ourselves through our body image, is key to the overall health and sustainability of individuals, societies, and Earth.

Being outside transforms relationships in many ways. Women are socialized to be relational and these authors postulate that helping women spend time outdoors in comfortable and safe-feeling environments will help transform the patriarchal culture through changes women may have with their relationships with their bodies and nature. As women become more comfortable with themselves and nature, they can return home ready to have agency and make changes. They can feel a renewed purpose in life; all of which have implications for personal, family, community, and environmental well-being.

Thus far the studies have demonstrated that spending time in natural areas likely contributes to a positive body image for women. There is likely an intersection of causes to both rebuff negative factors and to develop characteristics that may lead to women increasing their positive perception of their bodies.

Social comparison theory

Social comparison theory may still be in affect during time spent in nature. When women are part of an organized outdoor trip (as in two of the studies) or part of summer camp (as many study participants were) then the study participants may have been using the women guides or leaders who are likely comfortable in the outdoors and comfortable with themselves in a social comparison. Women outdoor leaders are physically capable and usually like their bodies and appreciate it for how it helps them function in the outdoors. Women may also do social comparison with nature and notice that nature is filled with beautiful imperfect beings. In this case they see many shapes and sizes of trees, other vegetation, and animals, learning that there is beauty in imperfection and that functionality is key.

Objectified body consciousness

Objectified body consciousness is possibly countered when in nature. Women tend to appreciate their bodies for what they can do and through that appreciation redefine what beauty looks like. What society may label as large thighs become 'thunder thighs' able to climb high mountains. As women relate to the natural area, they often appreciate and feel drawn to the natural world and therefore do not objectify it. They see many examples of vegetation in nature not able to control its appearance and the women see beauty in these trees and plants. This experience counters basic beliefs in objectified body consciousness that appearances can be controlled and that one's self worth and worth to others is determined by how we look. If women do not objectify themselves, it can have powerful repercussions when returning to their city or town environments. As women strengthen their body image, they may spend less time plucking, polishing, painting, toning, etc. themselves, resulting in more time and resources to use elsewhere. They may become more involved in environmental and other social actions, specifically because they now have a caring relationship with the natural environment.

Intrinsic health benefits of being in nature

Spending time in nature has been shown to have physical, psychological, spiritual, and emotional benefits, therefore some of the benefits women receive may be due to effects of the natural environment and some benefits may be due to being active. During time in nature, negative messages about women's bodies may be mitigated, and there is not the constant bombardment of media. When in natural environments, part of the rediscovery is understanding and feeling our sense of belonging. Many women travel in outdoor environments in order to feel a sense of coming or being home (Mitten 2010).

Positive body image

Wood-Barcalow and colleagues (2010) define positive body image as an overarching love and respect for the body that allows individuals to: (a) appreciate the unique beauty of their body and the functions that it performs for them; (b) accept and even admire their body, including those aspects that are inconsistent with idealized images; (c) feel beautiful, comfortable, confident, and happy with their body, which is often reflected as an outer radiance, or a 'glow'; (d) emphasize their body's assets rather than dwell on their imperfections; (e) have a mindful connection with their body's needs; and (f) interpret incoming information in a body-protective manner whereby most positive information is internalized and most negative information is rejected or reframed.

Combining the work of Wood-Barcalow and colleagues (2010) with what women report after participating on all-women outdoor trips, it seems that women often gain characteristics of positive body image while participating on these outdoor trips. Wood-Barcalow and colleagues (2010) reported that women who have an appreciation of shape, function, health, and features have positive body image. A major finding in the three body image studies reported about here was the appreciation of functionality of their bodies. Another finding from Wood-Barcalow and colleagues (2010) was that women with positive body image broadly conceptualized beauty and felt unconditional acceptance from significant others. As a metaphor in the more-than-human world, we see all sorts of shapes and sizes of trees and other vegetation, as well as animals. This signifies a level of acceptance and tolerance of differences; there are many ways to be beautiful. Many women commented that they felt at home or relaxed in nature. There is a feeling of acceptance from the natural world.

Wood-Barcalow and colleagues (2010) noted that women with positive body image tend to feel love and acceptance through their spirituality and by consciously choosing to associate with people who are accepting of themselves. Many women comment that they feel spiritually uplifted when in nature or that nature is their higher power. Fredrickson and Anderson (1999) reported that the women in their study appreciated the commonalities and said the group time helped them feel spiritual.

Often on outdoor trips women take care of their bodies by being physical, eating well, and sleeping well relating to the finding from Wood-Barcalow and colleagues (2010)—that women who have positive body image take care of their body via healthy behaviors. In the case of outdoor trips, the act of taking care of themselves could create a positive feedback loop about their body image that could carry into their city or town life.

Personal effectiveness

There is an important link between perceptions of personal physical effectiveness (increased through outdoor activities) and positive body image (increased when perceptions of personal effectiveness increase). A positive feedback loop can be created when women are pleased with their effectiveness—ability to be active in nature—and therefore feel positive about themselves and spend more time active outdoors.

Social and ecological impacts

The impact of a more positive body image through time in nature can have many implications for society. The act of spending time in nature is often seen as stepping out of prescribed societal norms. If a woman takes this step, she may take other steps outside norms. The act of healing oneself in nature gives one a closer relationship with nature. Women may advocate for environmental protection, clean water, and safe food sources. If women spend less time and money on fussing with their body, this time and the resources will go toward other endeavors. This might include more family time, creek cleanups, or other community service. The more positive women feel about themselves and the more they feel a sense of belonging, then the more we as a culture will go toward ecofeminism and a respect for all life.

Conclusions

An ecofeminist lens aids in our understanding about how the subjection of women and the environment are linked, as well as how encouraging a positive relationship with women and the environment can improve conditions for both. The consequences of a negative body image are far reaching—impacting many aspects of women's physical health as well as their emotional and psychological self-concept including self-esteem, self-efficacy, personal growth, and positive relationships. However, time in nature can be co-healing for both nature and women.

Newer research focusing on characteristics of positive body image provides information for clinicians and practitioner in health professions about directions for supporting women achieving and maintaining a positive body image. Research presented here strongly suggests that the positive body image characteristics described in this literature are promoted when women engage in outdoor activities on a regular basis.

Enabling women to achieve a positive body image can help women spend less time, energy and resources trying to achieve an unhealthy and unrealistic body ideal. Women are able to rebuff the patriarchal beliefs that contribute to diminishing and objectifying women and nature. Basically, some of the probable causes for a negative body image are moderated, including aspects of objectified body consciousness because women often stop monitoring how their body looks and thinking of themselves as an object to be looked at. While spending time in nature, women avoid the bombardment of media messages objectifying women, and if they engage in social comparison, they often have women who are comfortable with their body to which they can compare themselves. Women outdoor leaders, often comfortable in the outdoor environment as well as fit and able to engage competently with the outdoors, would be positive role models for other women to cue from in terms of lifestyle and self-acceptance.

Along with a decrease in factors that lead to negative body image, the characteristics of positive body image are reinforced, including broadly conceptualized beauty, and feelings of acceptance and of spiritual contentment while in the nature.

The study conducted by the chapter authors had two particular findings that have practical implications:

- Getting more than three to four hours of outdoor activity a week seems to make a significant difference in women's body image.
- Longer duration of outdoor activity in the form of number of concurrent overnights, total overnight trips, and total nights outdoors seems to have a positive impact on women's body image.

Results thus far reinforce the need to provide opportunities for women of all ages to engage in outdoor activities on a regular basis and to have opportunities during their lifespans to spend periods of time of one to two weeks outdoors. Young girls and their families can spend daily or weekly time together in nature or attend overnight camps as a family or individually. Women can be encouraged and supported in maintaining a lifestyle that includes regular time in nature.

Using these findings in this area of inquiry has the potential to be preventive for individuals and populations who may be at risk for poor body image and related issues. For example, young girls can be encouraged to play outdoors in order to help them create and maintain positive body image and using three hours as a dosage gives girls and caregivers a target.

Becoming aware of the physical and psychosocial variables surrounding body image may support the development of nature treatment programs for certain disorders, such as anorexia or body dysmorphic disorder. For women with negative body image, outdoor activities and time spent in the outdoors may help cause an increase in a woman's belief that her body is physically effective and that becoming physically effective feels good psychologically. Therefore a positive cycle could ensue of a woman spending time being active in the outdoors because it feels good, while receiving the many other health benefits from time in nature (Ewert et al. 2014), and perhaps not succumbing as easily to pressures to conform to unhealthy cultural standards of beauty. Women can return to their city or town life better able to engage in social and environmental actions.

Ecofeminism, or caring for all life, can be actualized through the benefits received from spending time in the natural world. The women may be able to forge a reparative relationship between women and the natural environment, leading to mutual healing. The combination of being physically active and feeling spiritually alive helps many women feel positive about their bodies and their being. Women who have been trapped in commodification and objectification just as nature has been commoditized and objectified may then find agency and come back renewed or transformed, ready to help transform the rest of the world to move toward more ecofeminist beliefs, which are inclusive of all and help everyone belong.

References

Avalos, L., Tylka, T. L., and Wood-Barcalow, N. 2005. "The Body Appreciation Scale: Development and Psychometric Evaluation," *Body Image* 2: 285–297.

Becker, A. E. 2004. "Television, Disordered Eating, and Young Women in Fiji: Negotiating Body Image and Identity During Rapid Social Change," *Culture, Medicine and Psychiatry* 28(4): 533–559.

Buddeberg-Fischer, B., Klaghofer, R., and Reed, V. 1999. "Associations between Body Weight, Psychiatric Disorders and Body Image in Female Adolescents," *Psychotherapy and Psychosomatics* 68(6): 325–332.

Cash, T. F. 1998. "The Emergence of Negative Body Images." In *Behavioral Medicine and Women: A Comprehensive Handbook*, ed. E. A. Blechman and K. D. Brownell. New York: Guilford Press.

Cash, T. F., and Pruzinsky, T. 2004. "Understanding Body Images: Historical and Contemporary Perspectives." In *Body Image: A Handbook of Theory, Research, and Clinical Practice*, ed. T. F. Cash and T. Pruzinsky. New York: Guilford Press.

Chawla, L. 2006. "Learning to Love the Natural World Enough to Protect It," *Barn* 2: 57–78.

Chiesura, A. 2004. "The Role of Urban Parks for the Sustainable City," *Landscape and Urban Planning* 68: 129–138.

Clarke, R. 1973. *Ellen Swallow, the Woman Who Founded Ecology*. Chicago: Follett.

Cleland, V., Crawford, D., Baur, L., Hume, C., Timperio, A., and Salmon, J. 2008. "A Prospective Examination of Children's Time Spent Outdoors, Objectively Measured Physical Activity and Overweight," *International Journal of Obesity* 32(11): 1685–1693.

Clements, R. 2004. "An Investigation of the Status of Outdoor Play," *Contemporary Issues in Early Childhood* 5(1): 68–80.

Coholic, D. 2011. "Exploring the Feasibility and Benefits of Arts Based Mindfulness-Based Practices with Young People in Need: Aiming to Improve Aspects of Self-Awareness and Resilience," *Child Youth Care Forum*, 40(4): 303–317.

Coker, E., and Abraham, S. 2014. "Body Weight Dissatisfaction: A Comparison of Women with and without Eating Disorders," *Eating Behaviors* 15(3): 453–459.

Collins, Patricia Hill. 1993. "Toward a New Vision: Race, Class, and Gender as Categories of Analysis and Connection." *Race, Sex & Class* 1(1): 25–45.

d'Eaubonne, F. 1974. "Feminism or Death." In *New French Feminisms: An Anthology*, ed. E. Marks and I. de Courtivron. Amherst: University of Massachusetts Press.

Dyl, J., Kittler, J., Phillips, K. A., and Hunt, J. I. 2006. "Body Dysmorphic Disorder and Other Clinically Significant Body Image Concerns in Adolescent Psychiatric Inpatients: Prevalence and Clinical Characteristics," *Child Psychiatry and Human Development* 36(4): 369–382.

Ewert, A., Mitten, D., and Overholt, J. 2014. *Human Health and the Natural Environment*. Oxford: Cabri Press.

Fallon, A. 1990. "Culture in the Mirror: Sociocultural Determinants of Body Image." In *Body Images: Development, Deviance, and Change*, ed. T. F. Cash and T. Pruzinsky. New York: Guilford Press.

Festinger, L. 1954. "A Theory of Social Comparison Processes," *Human Relations* 7(2): 117–140.

Flinders, C. 2003. *Rebalancing the World: Why Women Belong and Men Compete and How to Restore the Ancient Equilibrium*. San Francisco: HarperCollins.

Forehand, M. R. 2001. "What We See Makes Us Who We Are: Priming Ethnic Self-Awareness and Advertising Response," *Journal of Marketing Research* 38: 336–348.

Fredrickson, B. L., and Roberts, T. A. 1997. "Objectification Theory," *Psychology of Women Quarterly* 21(2): 173–206.

Fredrickson, L. M., and Anderson, D. H. 1999. "A Qualitative Exploration of the Wilderness Experience as a Source of Spiritual Inspiration," *Journal of Environmental Psychology* 19(1): 21–39.

Garner, D. M. 2002. "Body Image and Anorexia Nervosa." In *Body Image: A Handbook of Theory, Research, and Clinical Practice*, ed. T. F. Cash and T. Pruzinsky. New York: Guilford Press.

Grogan, S. 1999. *Body Image: Understanding Body Dissatisfaction in Men, Women, and Children*. London: Routledge.

Harris, A. 2015. "The Green Fuse for Environmental Philosophy, Deep Ecology, Social Ecology, Eco-Feminism, Earth-Centered Spirituality." www.thegreenfuse.org/ecofem.htm

Heeren A., and Philippot, P. 2011. "Changes in Ruminative Thinking Mediate the Clinical Benefits of Mindfulness: Preliminary Findings," *Mindfulness* 2: 8–13.

Hornibrook, T., Brinkert, E., Parry, D., Seimens, R., Mitten, D., and Priest, S. 1997. "The Benefits and Motivations of All Women Outdoor Programs," *The Journal of Experiential Education* 20: 152–158.

Howell, Andrew J., Dopko, Raelyne L., Passmore, Holli-Anne, and Karen Buro. 2011. "Nature Connectedness: Associations with Well-Being and Mindfulness." *Personality and Individual Differences* 51(2): 166–171.

Hug, S., Hansmann, R., Monn, C., Krütli, P., and Seeland K. 2008. "Restorative Effects of Physical Activity in Forests and Indoor Settings," *International Journal of Fitness* 4(2): 25–38.

Kaiser, S. B. 1997. *The Social Psychology of Clothing: Symbolic Appearances in Context*, 2nd rev. edition. New York: Fairchild.

Katz, N. 2004. "Body Image Statistics-Dieting Statistics-Body Type Statistics." www.womensissues.about.com/cs/bodyimage/a/bodyimagestats.htm

Kiewa, J. 1996. "Body Satisfaction and Competence: Hand and Glove?" *Social Alternatives* 15: 7–10.

Kiewa, J. 2000. "Outdoor Adventure and Body Image: A Change in Focus." In *Body Stories: Research and Intimate Narratives on Women Transforming Body Image in Outdoor Adventure*, ed. L. West-Smith. Edgewood, KY: Adventurehaven Press.

Korpela, K., Ylen, M., Tyrvainen, L. and Silvennoinen, H. 2009. "Stability of Self-Reported Favourite Places and Place Attachment over a 10-month Period," *Journal of Environmental Psychology* 29: 95–100.

Kuo, Frances E., and William C. Sullivan. "Aggression and Violence in the Inner City: Effects of Environment via Mental Fatigue." *Environment and Behavior* 33(4): 543–571.

Laudanska-Krzeminska, I., and Bronikowski, M. 2013. "Relationships Among Body Satisfaction, Life Skills, and Physical Education in Polish Adolescent Girls," *The Global Journal of Health and Physical Education Pedagogy* (1): 57–67.

Lee, J., Tsunetsugu, Y., Takayama, N., Park, et al. 2014. "Influence of Forest Therapy on Cardiovascular Relaxation in Young Adults," *EvidenceBased Complementary and Alternative Medicine* 2: 1–7.

Li, Q., Morimoto, K., Kobayashi, M., et al. 2008. "A Forest Bathing Trip Increases Human Natural Killer Activity and Expression of Anti-Cancer Proteins in Female Subjects," *Journal of Biological Regulators and Homeostatic Agents* 22(1): 45–55.

Maine, M., and Kelly, J. 2005. *The Body Myth: Adult Women and the Pressure to Be Perfect.* Hoboken, NJ: John Wiley.

Matthiasdottir, E., Jonsson, S. H., and Kristjansson, A. L. 2012. "Body Weight Dissatisfaction in the Icelandic Adult Population: A Normative Discontent?" *The European Journal of Public Health* 22(1): 116–121.

Mayer, F., Frantz, C., Bruehlman-Senecal, E., and Dolliver, K. 2009. "Why is Nature Beneficial? The Role of Connectedness to Nature," *Environment and Behavior* 41(5): 607–643.

McDermott, L. 2004. "Exploring Intersections of Physicality and Female-Only Canoeing Experiences," *Leisure Studies* 32: 283–301.

McGraw, S. 2012. "Why Body Image Matters." www.alreadypretty.com/2012/05/why-body-image-matters.html

McKinley, N. M., and Hyde, J. S. 1996. "The Objectified Body Consciousness Scale Development and Validation," *Psychology of Women Quarterly* 20(2): 181–215.

Mitten, D. 1992. "Empowering Girls and Women in the Outdoors," *Journal of Physical Education, Recreation, and Dance* 63: 56–60.

——. 2010. "Friluftsliv and the Healing Power of Nature: The Need for Nature for Human Health, Development, and Wellbeing," *Norwegian Journal of Friluftsliv*. http://norwegianjournaloffriluftsliv.com/doc/122010.pdf

——. 2016. "Connections, Compassion, and Co-Healing: The Ecology of Relationship." In *Reimagining Sustainability in Precarious Times*, ed. K. Malone, T. Gray, and S. Truong. New York: Springer.

Mitten, D. and Woodruff, S. 2010. "The Impact of Short Term Adventure Experiences on Body Image Perceptions of Women over 40," *Journal of Experiential Education* 32(3): 322–326.

Morrison, T. G., Kalin, R., and Morrison, M. A. 2004. "Body-Image Evaluation and Body-Image Investment among Adolescents: A Test of Sociocultural and Social Comparison Theories," *Adolescence* 39(155): 571–592.

Nisbet, E., Zelenski, J., and Murphy, S. 2011. "Happiness Is in Our Nature: Exploring Nature Relatedness as a Contributor to Subjective Well-Being," *Journal of Happiness Studies* 12: 303–322.

Peat, C. M., Peyerl, N. L., Muehlenkamp, J. J. 2008. "Body Image and Eating Disorders in Older Adults: A Review," *Journal of General Psychology* 135: 343–358.

Pohl, S. L., Borrie, W. T., and Patterson, M. E. 2000. "Women, Wilderness, and Everyday Life: A Documentation of the Connection between Wilderness Recreation and Women's Everyday Lives," *Journal of Leisure Research* 32: 415–434.

Potwarka, L., Kaczynski, A., and Flack, A. 2008. "Places to Play: Association of Park Space and Facilities with Healthy Weight Status among Children," *Journal of Community Health* 33: 344–350.

Robinson, T. R. 2003. "Clothing Behavior, Body Cathexis, and Appearance Management of Women Enrolled in a Commercial Weight Loss Program." Dissertation, Virginia Polytechnic Institute and State University.

Rodin, J., Silberstein, L., and Striegel-Moore, R. 1985. "Women and Weight: A Normative Discontent." In *Psychology and Gender: Nebraska Symposium on Motivation*, ed. R. A. Dienstbier and T. B. Sonderegger. Lincoln: University of Nebraska Press.

Ryan, R., Weinstein, N., Bernstein, J., Brown, K., Mistretta, L., and Gagne, M. 2010. "Vitalizing Effects of Being Outdoors and in Nature," *Journal of Environmental Psychology* 30: 159–168.

Sinclair, S. L. 2006. "Object Lessons: A Theoretical and Empirical Study of Objectified Body Consciousness in Women," *Journal of Mental Health Counseling* 28(1): 48–68.

Spake, A. 2004. "Rethinking Weight," *U.S. News and World Report* 136: 50–56.

Stein, T. and Lee, M. 1995. "Managing Recreation Resources for Positive Outcomes: An Application of Benefits-Based Management," *Journal of Park and Recreation Administration* 13(2): 52–70.

Tiggemann, M. 2004. "Body Image across the Adult Life Span: Stability and Change," *Body Image* 1(1): 29–41.

Tylka, T. L., and Wood-Barcalow, N. L. 2015. "What is and What is not Positive Body Image? Conceptual Foundations and Construct Definition," *Body Image* 14: 118–129.

Walker, G., Hull, R., and Roggenbuck, J. 1998. "On-Site Optimal Experiences and Their Relationship to Off-Site Benefits," *Journal of Leisure Research* 30: 453–471.

Webster, J., and Tiggemann, M. 2003. "The Relationship between Women's Body Satisfaction and Self-Image across the Life Span: The Role of Cognitive Control," *The Journal of Genetic Psychology* 164(2): 241–252.

Weinstein, N., Przybylski, A., and Ryan, A. 2009. "Can Nature Make Us More Caring? Effects of Immersion in Nature on Intrinsic Aspirations and Generosity," *Personality and Social Psychology Bulletin* 35(10): 1315–1329.

West-Smith, L., ed. 2000. *Body Stories: Research and Intimate Narratives on Women Transforming Body Image in Outdoor Adventure.* Edgewood: Adventurehaven Press.

Wheeler, L., and Miyake, K. 1992. "Social Comparison in Everyday Life," *Journal of Personality and Social Psychology* 62(5): 760–773.

Whittington, A. 2006. "Girls in the Woods: Exploring the Impact of a Wilderness Program on Adolescent Girls' Constructions of Femininity," *Journal of Experiential Education* 28(3): 285–289.

Williams, E. F, Cash, T. F., and Santos, M. T. 2004. "Positive and Negative Body Image: Precursors, Correlates, and Consequences." Paper presented at the meeting of the Association for Advancement of Behavior Therapy.

Wood-Barcalow, N. L., Tylka, T. L., and Augustus-Horvath, C. L. 2010. "'But I Like My Body': Positive Body Image Characteristics and a Holistic Model for Young-Adult Women." *Body Image* 7(2): 106–116.

Woodruff, S. K. L. 2009. "The Impact of Short Term Adventure Experiences on Body Image Perceptions of Women over Forty." Prescott College.

7 Writing women into back-to-the-land: feminism, appropriation, and identity in the 1970s magazine *Country Women*

Valerie Padilla Carroll

During the 1970s, U.S. back-to-the-land movement proponents advocated what they saw as a radical break with the American mainstream: the rejection of modernity in favor of the rural 'simple life' of environmentalist smallholder self-sufficiency. Promoted in magazines like *Mother Earth News* as well as a variety of books, back-to-the-land offered itself as a solution to the perceived economic, political, environmental, and cultural problems of the times. Problematically, but rarely mentioned as such, this radical response still rested on hegemonic assumptions that perpetuate heteropatriarchal power, such as the promotion of binary gender roles and devalued female labor. Indeed, women were generally absent in back-to-the-land literature. However, the emergence of Women's Liberation and more radical feminist thought offered a chance to write women into back-to-the land through self-published texts. This chapter explores these attempts in the feminist back-to-the-land publication *Country Women* (1973–1979) along with the editors' compilation how-to book *Country Women: A Handbook for New Farmers* (Tetrault and Thomas 1976). At its peak, *Country Women* magazine had gained a readership of over 10,000 with the publication's heady mix of feminist politics and DIY articles for the female back-to-the-land farmer. By exploring the many female voices in the thirty-three newsletter run of *Country Women* as well as the how-to book, we can see how women, so often left out of the back-to-the-land narrative, attempted to write themselves into the movement not merely as participants but as activist critics seeking to dismantle heteropatriarchy and define new empowered identities. However, such analysis will also show how these feminists, even as they sought equality, remained complicit in the oppression of Others via a reinforcing nature and racial/ethnic appropriation. This story, though, is not only one of privileged women gaining freedom on the backs of Others. Instead a more complicated, if not disheartening, story of women and nature emerges from the pages where multiple voices come together, explore possibilities, and seek guidance even as they miss the opportunity for truly radical change.

Writing women into back-to-the-land

In the U.S., the back-to-the-land movement has roots in 19th-century nature writing, but did not blossom as a popular movement until in the early 20th century. Back-to-the-land literature promoted an emancipatory narrative selling the promise of liberation via the story of everyman smallholder self-sufficiency. At its core, the back-to-the-land movement was, as agricultural historian David B. Danbom explains, "rooted in a critique of modern urban-industrial society and that society's effect on human happiness and right living" (1991, 2). Such writings tend to follow a similar story arc. They begin with the problems—unemployment, urban pollution, impure and unhealthy food, and a general discontent. The root causes vary: modernity, political corruption, industrialism, consumerism, or immorality. Yet the solution prescribed is the same: a retreat to the countryside where an agrarian self-sufficiency of gardening, animal husbandry, and self-built shelters offers meaningful work, autonomy, and security. These anti-urban narratives were an "over-arching antimodernist appeal to middle-class urbanites concerned with unchecked industrial growth, urban sprawl, and the fast pace of modern living" (Potts 2011, 820). Thus, these retreats to the countryside were a potent fantasy for many and were sold to the public as such. From the beginnings of the back-to-the-land movement, in books like progressive activist Bolton Hall's *A Little Land and Living* (1908), specialty magazines like *Country Life*, and more general magazines like *Colliers*, back-to-the-land was written as idyllic narratives wrapped in a "how-to" of self-sufficient agrarian living.

Such fantasies of independence and autonomy were, as Dona Brown (2011) explains in her comprehensive book, *Back-to-the-Land: The Enduring Dream of Self-Sufficiency in Modern America*, promoted as utterly masculine endeavors. Books and magazines alike promoted self-sufficiency as a method that would reaffirm and rebuild manhood and the traditional heteropatriarchal family that was perceived as threatened by modern, urban industrial life. Indeed, most early 20th-century back-to-the-land proponents rarely included women beyond the helpmate category. In the rare instances of female characters, like David Grayson's sister Harriet in his incredibly popular *Adventures in Contentment* (1906) or Ralph Borsodi's wife in *Flight from the City* (1933), the woman reads merely as the domestic labor support for his self-sufficiency adventure. Even so, women too sought the "very independence and self-sufficiency—the "manhood"—reformers thought they were safeguarding for men" (Brown 2011, 11).

By the 1970s, the back-to-the-land movement continued to rely on earlier 20th-century constructions. Typifying this focus can be seen in the flagship back-to-the-land magazine *Mother Earth News* where, in the inaugural issue, the editors assume a male audience and used the language of heteromasculinity to describe back-to-the-land asking, "is it actually possible to tell the boss to shove it, square your shoulders, and step out a free man?" (quoted in Brown 2011, 211). When women were included, the primary prescription offered naturalized gender roles. For example, in "Homesteading and Women's Empowerment", author Caroline Robinson (1970) unhesitatingly promotes naturalized binary gender roles when writing her follow-up letter to a series of articles written by her husband Ed Robinson. While Ed was careful to use the pronoun "they" to refer to his family's "have-more" plan for an urban retreat into the countryside, Caroline focused on defining gender roles when she wrote, "Mr. Robinson naturally does the heavy work in the garden and with the animals, while I take care of canning, freezing and household jobs" (Robinson 1970). Even when she assumes the language of Women's Liberation such as "women's empowerment," she equates support as empowerment, advising, "Your husband can't "Have-More" alone – he needs all your interest and help. But isn't that the way you want it?" (ibid.).

For the female back-to-the-landers seeking more than a helpmate role, advocating for women as homesteaders was aided by the nascent Women's Liberation movement as well as more radical feminist thought (Brown 2011, 212). *Country Women* as a feminist magazine emerged as an important forum for women seeking participation and visibility for themselves in back-to-the-land. Self-published out of Albion, California, the collectively written and edited journal covered a variety of feminist and back-to-the-land topics in its thirty-three-issue run. Half of each issue focused on practical DIY articles providing step-by-step instructions on a variety of subjects like chicken coop building, goat vaccination, and plumbing. Many of these articles along with some new ones were collected in the volume *Country Women: A Handbook for New Farmers* (Tetrault and Thomas 1976) offering a practical, how-to resource for all back-to-the-landers. The other half of each issue was devoted to a single theme, such as "Homesteading" (March 1973), "Food" (July 1976), and "Animals" (December 1977) and counter cultural explorations such as "Living Alternatives" (June 1973) and "Spirituality" (April 1974). Themes also included straightforward feminist subjects like "Women's Movement in the Country" (October 1973), "Women Working" (June 1975), and "Politics" (January 1976). Finally, other issues offered more radical and inclusive explorations such as "Children's Liberation" (August 1974), "Older Women" (July 1974), and "Sexuality" (April 1975).

While little scholarly work has been written on *Country Women*, Scott Herring's "Out of the Closet, into the Woods: 'RFD,' 'Country Women,' and the Post-Stonewall Emergence of Queer Anti-Urbanism" (2007) provides an excellent exploration of *Country Women*'s "critical rusticity" visual culture (359). Herring connects the lesbian separatist politics of the journal to the rural gay magazine *RFD* and its rejection of the metronormativity common in other gay liberation magazines of the 1970s. While Herring reads *Country Women* as lesbian separatism referring to its explicit promotion, it is important to remember separatism was expressed as part of a multiplicity of possible feminist identities and practices. Indeed, the manifesto printed in the first four issues offers the publication as a resource for all women:

> We see Country♀Women as a feminist country survival manual and a creative journal. It is for women living with women, with men, and alone,

> for women who live in the country already and for women who want to move out of the cities. We need to learn all that women can do in the country and learn to break out of oppressive roles and images. We need to reach out of our isolation from one another, to know that we aren't alone, that we aren't crazy, that there is a lot of love and strength and growing to share. Country♀Women can bring us together.
>
> (1973, 1)

Consequentially then, while *Country Women* did promote lesbian separatism, it was offered as one of many methods of centering women's relationships. The female-to-female relationship was promoted as a feminist practice whether manifesting as life partners, lovers, or as friends. Because *Country Women* was aspirationally inclusive, "for women living with women, with men, and alone, for women who live in the country already and for women who want to move out of the cities," it could house multiple voices with multiple perspectives centered around feminist issues of back-to-the-land. It was a site where women could help women and develop community. Further, because *Country Women* was composed of a multiplicity of authors, editors, and artists, each voice contributed to an ongoing dialogue seeking to define new social and cultural roles, values, and identities.

Country Women provided much important feminist and female-centric information and analysis for women back-to-the-landers, including the promotion of body positivity, step-by-step instructions on self-defense, and open and frank discussions of sexual assault and domestic violence. In ways both similar and different to Adriana Teodorescu's assertion that "women have made a great effort to control motherhood, separate it as much as possible from natural laws and disrupt the connection between women and nature" (Chapter 5 this volume), *Country Women* offered information on reproduction that at times separated motherhood from nature and at other times defined birth control and reproductive choice as natural. Some of the strongest feminist analyses though attacked the entrenched heteropatriarchy both inside and outside the back-to-the-land movement and its literature. For example, articles like "Building a New Foundation" (Volkhausen 1975, 7–8) questioned back-to-the-land couples' unthinking dependence on gendered labor, while "Housewifery: A Historical Perspective" (Curtis and Bye 1975, 18–19) explored the ways that housework, while vital, need not be gendered.

Even the DIY articles provided astute feminist critiques intertwined with the instructions. Next to directions for the maintenance and use of chainsaws, for example, the power saw is written as, "a sacred male tool" that transforms in the hands of women to become a radical act of resistance and empowerment (Ross 1976, 72). Essentially, the confident use of tools was seen as a feminist act that challenged masculine power and allowed women to develop as self-sufficient individuals. In another example, author Sherry Thomas explains that after learning to build a shed, "I stood in the middle of that dirt-floored, low-ceilinged, plywood shed and knew that where-ever I went after that, I would always be able to provide myself with shelter. And since then I have had a new sense of possibility fear does not paralyze me anymore when I know I need a lean-to for milking or a cabinet to hold my tools" (1972, 25). Sally Bailey in "Fisherwomen" (1972) explains the same feeling by connecting the learning of masculine skills to personal empowerment, "Learning new skills and being back on the ocean I gained a new self-respect" (20). Harlene Amberschild (1976) explains her confidence grows with the use of tools, "As a woman, I feel especially good about learning how to work with the land and with tools and machines. I feel good about my body growing strong along with my confidence to do things necessary to my existence" (19). Such feminism in the practical skills section of each magazine and in the compilation is important because how-to can often appear apolitical. Yet these step-by-step instructions in *Country Women* offered access to a hidden world of masculine skills and information thereby providing fertile ground for feminist analysis of female labor and tool use outside of gendered roles.

Some of the most forward thinking feminist work in *Country Women* was a kind of intersectional analysis and critique of class and gender connected to capitalism. While covered in many of the magazine's issues, the best examples come from February 1977's aptly titled theme issue, "Class." From the beginning of the issue, critiques of social position and the promotion of lived knowledges were in the forefront. Importantly, the editorial collective named their social position and potential bias before embarking on their discussion of class and gender. Much like Kimberly Crenshaw's work on intersectionality a decade or so later, they "name their perspective" with a clear statement: "This magazine is produced by predominantly white, middle class women" (Crenshaw 1991, 1244; Collective 1977, 1). More so, they recognized their position of ignorance and the process of self-education when they wrote, "The consciousness raising, the reading, the interpersonal struggles, and the soliciting of material culminate in a magazine before a lot of the new ideas and feelings are digested and/or integrated into our lives. Our struggle with class is a continual process" (ibid). While the editors may have been unsure of their place in the class struggles, the article submissions clearly came from a site of lived knowledge coupled with astute socio-economic analysis. This included, for instance, a woman's personal experience tied with an exploration of how socio-economic position impacts the family relationships and interpersonal violence (Ashe 1977, 6). Another article considered, from a particularly Marxist stance, how class was tied to "the amount of vested interest we have in maintaining the present order, the domination by the tiny ruling class" (together-almost-a-year study group 1977, 2). In a rare focus on the intersections of gender, class, and race, the article "Land as a Means of Production" explored one woman's experience with land ownership as power as well as the historical and contemporary ways that women and people of color have been both denied land ownership and had their labor appropriated in the interest of the ruling class (Terry 1977, 28–32). Indeed, outright calls for solidarity across class, race, and nation can be found in "Target: Capitalism" which calls for "international solidarity," explaining, "in the struggle to over-throw ruling class oppression, working classes throughout the world share common bonds. It is necessary to understand how the capitalist system exploits people at home and abroad" (Weed et al. 1977, 17).

As a whole then, *Country Women* offered women a way to write themselves into self-sufficiency, grow as an individual, fight for gender justice, and develop strategies of change. It also provided some real proto-intersectional analysis by questioning the connections between gender and classism and gender and imperialism even calling for the dismantlement of capitalist hegemony. Such solidarity often tried to avoid homogenization as the women still recognized (some) difference, especially in class and sexual orientation. Such multiple and even oppositional perspectives can be a vital to the evolution of a movement. It is here, however, that I return to the complicated and disheartening story promoted by *Country Women*. Even as they sought equality and recognized the importance of class position bias, other oppressive constructions remained unquestioned. The aspirational inclusiveness of these feminists rarely extended to nature or even to people of color who are stereotyped as more "natural." Instead nature and people of color are used as tools for the development of the individual white feminist self. That is, like the more masculine back-to-the-land movement they were writing women into, *Country Women* still maintained an unquestioned acceptance of the primacy of the individual, in their case the white Western female back-to-the-lander.

Othering others: the ecological Indian and nature as a tool for identity

Otherness not only constructs particular groups as marginalized but is also based on an unquestioned hierarchical assumption that locates and defines the powerful as standard and the less powerful as outsider. *Country Women* was in many ways a response to the Othering of women in back-to-the-land. By questioning and exposing heteropatriarchal assumptions and power, these feminists demanded to be seen and treated as full and equal human beings. They went beyond just writing women into the back-to-the-land movement—*Country Women* was feminist activism and resistance. Nevertheless, coming from race, global location, and species privilege, many of the writers, editors, and artists in *Country Women* also othered Others by appropriating both real and imagined non-Western and non-white cultures as well as nature via the trope of the ecological Indian and a human/nature hierarchical binary. In fact, both were constructed as tools for the development of the Western white female identity.

Throughout *Country Women* there were numerous articles, poems, stories, and images, which used people of color as tools of feminist self-empowerment. For example, there are Western interpretations and selective use of Eastern mysticism including yoga and t'ai chi as well as the culturally unmoored use of icons like the yin/yang or the symbol that, "helps us get into who we are right now and lets us see our own preciousness, helping us to accept ourselves as whole worthy persons" (Purple 1974, 2). Such interpretations remove the practices and symbols from their cultural context and are reinterpreted to the benefit of the Western individual, in this case the female back-to-the-lander seeking to develop self-awareness and self-acceptance. Additional ways that *Country Women* Othered non-Western cultures and people was to fetishize them as uncontaminated by civilization. Hence non-Western people were claimed to have extraordinary human abilities because they are uncorrupted by Western civilization such as in "A Holistic Approach to Contraceptives" which claimed that in India there is a, "seemingly fantastic ability among the unmarried who use mind control as a contraceptive" (Kernis 1977, 52). Indeed, the author explains, the non-Westerner is an almost supernatural object in relation to the Western subject:

> When we read anthropological studies of tribal people who demonstrate outstanding acuity in hearing or smelling over great distances, we are surprised that the human ability is so vast. Then we begin to realize the limitations we have placed upon ourselves.
>
> (Ibid.)

Thus, the Westerner has much to learn from "tribal people" who can free the Western subject from "the limitations we have placed upon ourselves." While such descriptions seem positive, they exact boundaries separating the us or "ourselves" from the generalized Others, as the article says, "we" learn from "anthropological studies of tribal peoples." Aside from this generalized Othering, such descriptions also point to a major theme throughout the magazine's run: the idea that non-Western and non-white peoples are tools for use by the white subject.

More than any other people of color used as tools, the Native Americans, especially the imagined Native Americans, are offered repeatedly as the guides for self-empowerment and even redemption of whiteness. As Dakota scholar Phil Deloria explains in his foundation work, *Playing Indian* (1998), from the early days of European encounters with the Americas, indigeneity has been constructed and appropriated to represent and build the white American identity. The 1970s back-to-the-landers were no different, simply following "their cultural ancestors in playing Indian to find reassuring identities in a world seemingly out of control" (Deloria 1998, 158). Like other countercultural and even mainstream environmental groups in the 1970s, the activists of *Country Women* looked to the "ecological Indian" for guidance. This imagined Native person embodied the supposed inherent ecological sensibility of Indigenous peoples. He is a trope constructed as a wise teacher who understands "the systemic consequences of his actions, [he] feels deep sympathy with all living forms, and takes steps to conserve so that earth's harmonies are never imbalanced and resources never in doubt" (Krech 2000, 21). According to ecofeminist Noël Sturgeon (2009), the ecological Indian trope had its "decisive entry . . . onto the contemporary popular cultural stage" in the famous early 1970s anti-littering television advertisement where "Indian" actor Iron Eyes Cody shed a single tear at the littering and pollution of modern America (65). Sturgeon continue, "The powerful image of the crying Indian convinced many that littering was wrong, that environmentalism was important (and not too radical), and, not incidentally, that Indians were natural ecologists" (ibid.).

In *Country Women* a similar deployment of the ecological Indian was used to condemn aspects of Western culture that the feminists named imperialist aggressions. For example, the article, "Would You Like to Buy Rhode Island?" explained, "Indians lived on this very ground centuries without claiming ownership of it. The land 'belonged' to those who used its plants, animals, streams, wind, until all were displaced by white men's greed" (Thomas 1973, 4). Here is an acknowledgment of the imperialism and oppressive nature of Western culture, the wrongdoing (attached to 'white men') which, it is implied, can be cleansed by learning from the wise Native who can offer connection to the land and a better way of life. Even more explicitly in its condemnation of imperialism and its use of the ecological Indian as a tool, Pelican in "Purification in the Nuclear Age" (1979) calls for "us" to be aware of "Native American teachings and prophecy because there is a message in it for all people about the future of the earth and how to proceed. The traditional Native American cultures are living examples of a practical alternative to the suicidal white man's Western Civilization" (20). Here traditional Native American cultures like the Hopi are seen to offer guidance to whites. Their prophecies are for all people, white people included, and they offer examples of a practical alternative to the "suicidal white man's Western Civilization." Pelican continues:

> We need to look to Native people's traditional way of life and values for guidance and inspiration. They are the original people of this continent. Guided by their spiritual principles, they have lived a way of life that is in harmony with Mother Earth for a very, very long time. Ancient spiritual-based communities in harmony with Nature and Mother Earth, such as the Hopi, must especially be protected, and not forced to abandon their way of life and the natural resources which their lives depend on. The rights of the traditional peoples to survive as a people must be recognized and supported. The policies of cultural and physical genocide must be stopped, by first stopping all further uranium and other mining, and industrial development, on their land
>
> (23)

In this paragraph we can see how the complications of appropriation and solidarity intersect. Pelican deploys the ecological Indian, the idea of Indigenous oneness with the land, "living in harmony with mother earth" and she includes a moment of social justice solidarity, calling for the end of uranium mining on Native lands for example. Yet that call for solidarity is not exactly in the service of the Native peoples but of critical importance to the future of white people, for example, "We need to look to" the Hopi. The "we" in this statement is the white Western subject; the Native is the Other who is offered as a tool.

Another way that indigeneity gets deployed in *Country Women* is through what Deloria explains as the misuse of " symbolic Indianness," which includes the imagined and real artifacts and symbols of Native peoples (1998, 154). The feminist back-to-the-landers in *Country Women* appropriated these imagined artifacts and symbols of indigenousness to construct self-identity. Thus, Medicine Wheels, sun dances, and totem poles provided concrete expressions of enacting the guidance of imagined Indigenous cultures (Jacobs et al. 1974, 7–10; Goodyear 1973, 32–33; Woodsorrel 1974, 51–55). Above all, it was the tipi that was the most common cultural artifact that various *Country Women* authors appropriated. The tipi was seen as a kind of Native spiritual guide, a physical representation of the ecological Indian. As an illustration, in the article "Consider the Circle" an author named River (1973) explored the ways the tipi itself acted to help her discover the truth of life and nature. She wrote, "Knowing deep in my White Anglo Saxon Protestant bones that the tipi itself, the very structure I'd hastily chosen, was acting on my psyche, my spirit, as no structure ever had before" (28). She continued to discuss how the tipi and by extension "Indians" allowed her to release the dead weight of mainstream whiteness and embrace a spiritual connection with the earth:

> After living a short while in my tipi as a totally unreconstructed white woman, I began to feel the heavy pulls of heritage and history. The two wooden chests I'd moved in for storage of clothing and treasure seemed pompous, ungainly, un-Indian, and they were. I moved them out again. Kept a small pile of garments near my bed, put possessions in a few Indian baskets I'd happily inherited, and sat on a rug, on a tarp, on the earth. On the earth! A new, bright astonishment at how far I've lived from the earth in my life.
>
> (Ibid.)

Here River connects to the earth and meaning through the tipi, which functions as her spiritual teacher who influences her growth as a self. This ideal lifestyle where she finds meaning is something she sees as inherited and rightfully owned. While River seems to be admiring "Indian-ness," such admiration manifests as romantic conceptions divorced from history and actual ethnic cultures. Gone is the colonization and genocide. Instead, like the "Indian baskets she happily inherited," this idealized culture is employed for the individual woman's self-empowerment and to alleviate *her* "heavy pulls of heritage and history" of mainstream whiteness. Such cultural appropriation of the Other, as bell hooks reminds us, "assuages feelings of deprivation and lack that assault the psyches of radical white youth who choose to be disloyal to western civilization" (1992, 26). River's disloyalty to mainstream whiteness has her searching for another identity neither Native nor white but newly constructed from bits and pieces of the Other.

Overall the cultural appropriation of Indigenous artifacts and the definition of Indigenous peoples as teachers for the development and empowerment of white women permeated the magazine. For the authors in *Country Women*, the ecological Indian with his imagined deep connection to the earth offers guidance to young white women. In other words, through the deployment of the ecological Indian trope, *Country Women* presented Indigenous Americans and Indigenous cultures as tools for the white soul. The irony of this appropriation is that these feminists could easily recognize and condemn women as oppressors. They even produced entire themed issues that recognized and condemned female as oppressor showing the ways that, for example, middle-class women could oppress lower-class women and how straight women could oppress lesbians (*Country Women: Class* 1977; *Country Women: Sexuality* 1975). They even acknowledge the ways that the mother role could oppress children (*Country Women: Children's Liberation* 1974). Yet at the heart of these realizations, the "women" in *Country Women* were still coded as white Western woman. Entrenched in the privileges of American whiteness, these appropriations remained unquestioned even presented as sincere admiration for such cultures. It is the apparent sincerity of these women that points to the structural nature of oppression—it is so much broader, deeper, and more intertwined into culture that even when these feminists called for radical change, the power of hegemonic control remained.

The appropriation of nature by the feminists of *Country Women* is a little more complicated because the human/nature binary was both challenged and naturalized. As an environmentalist magazine, *Country Women* rejected the mainstream commodification of nature, animals, and the land, or those ecofeminists Karen Warren calls "earth others" (2000, 1, 125). Even as they critiqued the capitalist patriarchal misuse of earth others, however, *Country Women* still used nature as a resource for the empowerment of the white, Western subject. Earth others are presented as the teachers and healers for the individual human self. However, existing in tension with this anthropocentrism was a deep desire to seek connection and alliance by defining nature as both a sacred goddess and a feminist project.

The most obvious appropriation of earth others in *Country Women* were, like the ecological Indian, constructing the land, plants, and animals as teachers and healers for these women. Sometimes it is stated outright such as when as when Barbara from Covelo proclaims in "An Alternative to the Homestead Approach," "The land was, and is, a healer for me" (1973, 2). Sherry Thomas in "Being is Believing" (1977) explains gardening as a journey of self where the garden "has become my teacher and healer" (7). More often though nature is used as an analogy such as in "Cycles: Reflections" where the experiences on the farm are explained as, "I am learning to grow here. I'm learning to grow from the tomato . . . I am learning how to grow here. And I am learning how to die here" (Jacobs et al. 1974, 7). Here the author lyrically emotes growing tomatoes as a connection to the earth, to life, and to death. The tomato is a teacher that mirrors the author's life helping her understand her own mortality. Like Mitten and D'Amore discuss in "The Relationship of Women's Body Image and Experience in Nature" (Chapter 6 this volume), the authors in

Country Women understand the positive health benefits of being in and connecting to nature. However, these authors also define nature as more than a site for developing physical and mental health, they anthropomorphize nature as a teacher and healer as a means of empowerment for the human subject. In other words, the women grow as autonomous subjects whereas nature is an object that provides the resources for that growth.

Sometimes in tension, other times in conjunction with using earth others as teachers, *Country Women* also offered a metaphorical merging of women and earth others, most often in the funky artwork that dotted the pages. For instance, a drawing in the themed issue "Spirituality" shows a woman with arms raised to the sky. Naked to the waist, she is clad only in a skirt woven out of whole, living, earth others—a multitude of animals and plants including flowers, lizards, vines, camels, giraffes, snakes, pigs, horses, octopus, and butterflies among others (Guignon 1974, 13). In the same issue, another illustration depicts a pregnant woman as a tree, her legs rooted in the earth, her hair and arms as branches reaching toward the sky (Flores 1974, 21). Another manifestation of the woman/earth merging was through the trope of the goddess, often as a personified Mother Earth. Particularly in the early years of the magazine's run, this goddess image appeared repeatedly in artwork promoting a conflation of the female body and the land. For example, accompanying a poem about the first Women's Country Festival (1973) is a line drawing of a supine young female as a mountainous landscape—her breasts, belly, and knees forming the outlines of mountains (Leona 1973, 15). In another issue, a drawing titled "Mother Earth Lain to Rest" presents the landscape as a headless woman's body where her breasts and abdomen shapes mountains and her pubic hair forms the trees (Koolish 1973, 6). In yet another illustration, the connection of goddess and woman is shown in a six-pane drawing reminiscent of a comic book page. In each window, nature is anthropomorphized as female body parts. The sun is woman's face weeping, her hair streaming down as lightning. In another pane a woman with her cheeks puffed as she blows out water complete with fish. Another window shows a mountain where between vaginal folds flows a waterfall (Young 1974, 3). Such images show women, nature, the earth, and goddess as transposable and are meant to recognize a deep and meaningful, often essentialist connection between women and nature. These drawings evoke the goddess, a concept that theologian Carol Christ (2012) explains as, an "affirmation of female power" and a "positive valuing of female will" (249, 248). Such images attempt a liberatory message where women's power derives from a perceived innate connection to the earth outside of patriarchal culture.

While the artists may have hoped to evoke female connection to and power with earth and earth others, the poem next to Young's six-pane drawing returns the idea of earth/goddess as teacher and while simultaneously reinforcing the separation between human and nature by claiming to speak for the earth. In this poem, Ruth of Mountain Grove writes of her connection to the goddess: "I have dedicated my body / To the goddess / Now I must wait / To learn what will be, / To know, to know" (1974, 3). As her teacher, the goddess will offer direction and purpose. Yet she ends the poem speaking for the earth, "I will bring through my words Your message / Your message; / your voice / Into a woman's world" (ibid.). Here the woman is both student and teacher drawing from the goddess and speaking for her to create a "woman's world." What begins as the spiritual growth of the woman becomes a feminist political project for the benefit of all women.

Such spiritual conflation of women and nature happens less as the magazine matures through the 1970s. Throughout all the issues though, but especially in later ones, the connection of women to nature was written as an explicit political project where the enlightened feminists are portrayed as the protectors and spokespersons for earth others. Unlike the goddess approach, this one defined the women as similar but separate from nature. Thus, the relationship between women and nature was not about sameness but rested on the feminist duty, even *noblesse oblige*, to protect nature from the destructive forces of capitalist patriarchy. For example, in "No Trespassing," author Carmen Goodyear sees herself as obligated to protect the earth others on her land from the unenlightened population even if it brings, "'bad vibes' from people who want to use the land but not care for it" (1973, 25). Echoing this obligation Diney Woodsorrel in "A Land Buying Guide" explains that "I feel one of the biggest reasons for we land-loving people and especially women to take this [the protection of nature] on is to combat the old male trend of heavy developing and misusing and raping and turn it toward care and furthering life and happiness" (1973, 10). Here the "especially women" is telling. She is describing a particular obligation based on a feminist identity that requires the dismantlement of patriarchy or the "old male trend" of destructive land abuse.

Connecting these women's obligations to earth others, especially the land, has at its foundation an inkling what ecofeminist Karen Warren explains as the "one central conceptual issue [that] concerns the nature of the interconnections, at least in Western societies, between the unjustified domination of women and 'other human Others,' on the one hand, and the unjustified domination of non-human nature, on the other hand" (Warren 2000, xiv). While *Country Women* rarely recognized their similar oppressive experiences with other human Others like Indigenous peoples, throughout the magazine there is a trend toward recognizing the ways that women and earth others are similarly oppressed. Put most clearly by Linda Ford in "Journey" (1977), she discusses the connection between feminist anger and "the needs of the planet, our home, our Mother" (29). She continues:

> Maybe the timing is such that women have found their voice and their indignation and fury about the personal levels of oppression and hurt that they have experienced as women, and with that practice under our belts, the time has come when we can fulfill our role as spokespersons for the Earth Mother, with whom our experience as women in this lifetime gives us identity.
>
> (Ibid.)

Ford wants feminists to channel their anger, "their indignation and fury" for the protection of the earth. Here the theory and practice of these women's feminism recognizes similarities in oppression and even in identity between that women and the earth. This is not a spiritual merging of woman and nature but a feminist fire born of women's liberation and radical feminist thought where women "found their voice" and got "practice under their belts." Yet, in the end, the journey was too far, returning to the language of separation, Ford sees this connection between women and nature as a site where women speak for earth others, even more so, the enlightened feminist is the "spokesperson for the Earth Mother." She creates a new role for women that removes the agency of nature even as that nature gives women identity.

Throughout *Country Women*, whether using earth others as teachers, merging symbolically with nature, or as spokesperson for the earth, at the foundation of these arguments is the unquestioned primacy of the white Western subject. Earth others and other human Others are constructed in these pages as tools for personal growth. Even as political projects, such as being the spokesperson for the earth, liberatory intention rests on the individual purpose of the feminist activist. As a feminist magazine they consistently explored and condemned the ways that women were oppressed and made invisible by American culture. They recognized the oppressions of earth Others wrought by patriarchal capitalist power and promoted an activist response to protect of earth Others. The recognition of similar experiences of women and nature offered an opportunity for these women to fully relate to and work with earth Others. Unfortunately it is an opportunity missed since the calls quickly devolved into speaking for nature as a savior rather than as an equal. These women, these feminists, these radicals, desperately wanted to create egalitarian spaces, yet unfortunately they remained mired in a white colonial (and speciesist) framework that demanded they other Others even as they themselves were gender Others.

I close by describing where most of the readers of *Country Women* would start: the cover. *Country Women*'s cover for twenty-six of their thirty-three issues was the same line drawing consisting of a central title encircling a raised fist feminist power symbol.

Held firmly in the fist is a cluster of wheat. In opposite corners of the cover illustration are four portraitures of different women. These figures are outlines, drawn without facial features. One woman sows seed from a satchel, another reaps by swinging a scythe, and another woman picks apples, while the final woman sits reading by a wood stove. Bordering the cover, touching but never entering the portraits, are vines of cucumber, pumpkin, grapes, and beans. The hypothetical reader, a women in the 1970s, perhaps on the land, perhaps just dreaming, begins with the cover. The cover shows the possibilities and the promise of back-to-the-land for women. Each woman has agency in working the land whether in sowing, reaping, harvesting, or even enjoying the fruits of one's labor relaxing by a fire. These are independent, feminist women. Such imagery speaks to the heart of the desires of many women seeking the empowerment and autonomy that comes from self-sufficiency. Yet this cover illustration also points to a foundational ideological problem of in *Country Women*, perhaps even the back-to-the-land movement. The unquestioned primacy of the individual defines the self-sufficient back-to-the-lander as independent and autonomous. Conversely, a feminist activism demands collectivity, collaboration, and integration. This tension between the desires of individual autonomy and feminist collective sisterhood permeated the entire magazine's run. Herring, again in "Out of the Closets into the Woods," reads *Country Women*'s cover as connective stating, "Though each female is singular, these women are braided together through an illustrated border of grape and pumpkin vines" (2007, 356, 358). However, my reading is not so hopeful. The artwork represents what I believe to be the central obstacle to a truly radical advocacy connecting women and earth others (and with other human Others). Separate, faceless, and coded as white or at least not explicitly raced, these women seek connection but never truly achieve it. In the image, as in the writings contained within, the portraits remain individual, even autonomous, with each woman separate from each other and connected to nature only in how they work or use the land.

References

Amberschild, Harlene. 1976. "Struggle in the Country." *Country Women: City-Country*. December.

Ashe. 1977. "Ashe Speaks." *Country Women: Class*. February.

Bailey, Sally. 1972. "Fisherwomen." *Country Women: Work and Money* 4: 20.

Barbara, Carol, Kathy, Tania of Covelo. 1973. "An Alternative to the Homestead Approach." *Country Women: Living Alternatives*. June.

Brown, Dona. 2011. *Back-to-the-land: The Enduring Dream of Self-Sufficiency in Modern America*. Madison: University of Wisconsin Press.

Christ, Carol. 2012. "Why Women, Men and Other Living Things Still need the Goddess: Remembering and Reflecting 35 Years Later," *Feminist Theology* 20(3): 242–255.

Collective. 1977. "Opening Statement." *Country Women: Class*. February.

Crenshaw, Kimberly. 1991. "Mapping the Margins: Intersectionality, Identity Politics, and Violence against Women of Color," *Stanford Law Review* 43(6): 1241–1299.

Curtis, Nancy and Harriet Bye. 1975. "Housewifery: A Historical Perspective." *Country Women: Women Working*. June.

Danbom, David E. 1991. "Romantic Agrarianism in Twentieth-Century America." *Agricultural History* 65(4): 1–12.

Deloria, Phil. 1998. *Playing Indian*. New Haven: Yale University.

Flores, Alice. 1974. Untitled line drawing. *Country Women: Spirituality*. April.

Ford, Linda. 1977. "Journey." *Country Women: Anger and Violence*. September.

Goodyear, Carmen. 1973. "No Trespassing." *Country Women: Women and Land*. June.

Guignon, Valerie. 1974. Untitled line drawing. *Country Women: Spirituality*. April.

Herring, Scott. 2007. "Out of the Closet, into the Woods: 'RFD', 'Country Women', and the Post-Stonewall Emergence of Queer Anti-Urbanism," *American Quarterly* 59(2): 341–272.

hooks, bell. 1992. *Black Looks Race and Representation*. Boston: Southend Press.

Jacobs, Helen, Mahala Greenburg, Sherry Thomas, Carmen Goodyear, and Bobbi Jones. 1974. "Cycles: Reflections." *Country Women: Cycles*. December.

Kernis, Merilee. 1977. "A Holistic Approach to Contraception." *Country Women: Animals*. December.

Koolish, Linda. 1973. "Mother Earth Laid to Rest." *Country Women: Women and Art*. N.d.

Krech, Shepard. 2000. *The Ecological Indian: Myth and History*. New York: W. W. Norton.

Leona. 1973. Untitled line drawing. *Country Women: Women and Land*. June.

Pelican. 1979. "Purification in the Nuclear Age." *Country Women: Future Visions*. N.d.

Potts, Dale E. 2011. "Community within Nature: Culture and Environment in the Chimney Farm Literature of Henry Beston and Elizabeth Coatsworth, 1944–48," *Interdisciplinary Studies in Literature and Environment* 18(4): 820–839.

Purple, Marnie. 1974. "Biorhythms." *Country Women: Cycles*. December.

River. 1973. "Consider the Circle." *Country Women: Women and Land*. June, 28–29.

Robinson, Caroline. 1970. "Homesteading and Women's Empowerment." *Mother Earth News*. March/April. CD-ROM.

Ross, Jenny. 1976. "Chainsaws." In *Country Women: A Handbook for New Farmers*, ed. J. Tetrault and S. Thomas. Garden City: Anchor Books.

Ruth of Mountaingrove. 1974. "Poem." *Country Women: Spirituality*. April.

Sturgeon, Noël. 2009. *Environmentalism in Popular Culture*. Tuscon: University of Arizona Press.

Terry. 1977. "Land as A Means of Production." *Country Women: Class*. February.

Tetrault, Jeanne and Sherry Thomas. 1976. *Country Women: A Handbook for New Farmers*. Garden City: Anchor Books.

Thomas, Sherry. 1972. "Pole Framing." *Country Women: Homesteading*. March.

——. 1973. "Would You Like to Buy Rhode Island?" *Country Women: Women and Land*. June.

——. 1977. "Being is Believing." *Country Women: Personal Power*. April. together-almost-a-year study group. 1977. "Reflections in a Class Eye." *Country Women: Class*. February.

Volkhausen, Janetrose. 1975. "Building a New Foundation." *Country Women: Feminism and Relationships*. October.

Warren, Karen. 2000. *Ecofeminist Philosophy: A Western Perspective on What It Is and Why It Matters*. Lanham, MD: Rowman & Littlefield.

Weed, Zee, Julie, and Patsy. 1977. "Target: Capitalism." *Country Women: Class*. February.

Woodsorrel, Diney. 1973. "A Land Buying Guide." *Country Women: Women and Land*. June.

——. 1974. "Country Living Skills: Tipis." *Country Women: Spirituality*. April.

Young, N. C. 1974. Untitled line drawing. *Country Women: Spirituality*. April.

Part IV

Mediating practices

8 Bilha Givon as Sartre's "third party" in environmental dialogues

Shlomit Tamari

Bilha Givon is a well-known personality in the environmental movement in Israel. In the book *The Environment in Israel* (Tal 2006), there is a relatively detailed description of some of the environmental struggles in which she was involved. In this book she is defined as "one of the few women who managed to survive the predominantly male hierarchy of the Society for the Protection of Nature in Israel (SPNI), for a long period of time, and as one "who has achieved significant victories for the green movement." In 2006, the organization for the Sustainable Development of the Negev (SDN), which Givon established, received the Transparency Award on the Public Forums with industry (Transparency International–Israel); in 2008, Givon won the "Lifetime Achievement Award for Environmental Protection" (by the Minister for Environmental Protection and the President); in 2009, the SDN organization received the Knesset Speaker's award in the category of Quality of Life; in 2013, Givon was elected "Woman of the Year" in the south, on behalf of Lions Club International; in 2014, SDN won the Cleantech award.

This chapter will focus on the period when Givon was one of the leaders of the mediation process between the Ministry for Environmental Protection and the chemical plants in the Ramat Hovav industrial zone.[1] Beginning to collect material for this study, I was guided by the thought that Givon could not have been invited from the outside into the mediation process as an objective and impartial mediator. After years of struggle against the chemical pollution caused by the plants, after being considered the biggest enemy of the chemical industry in the Negev, why did she decide to take it upon herself to ease the regulator in his negotiations with the owners of capital?

Knowing the successful results of the mediation process and being familiar with Givon as well, I wanted to reflect with her on the process she underwent between 1998 and 2008. What existential questions was she dealing with, what processes had she to undergo to change from a darling of the SPNI, to a woman who for years was seen as having crossed the line and as someone who has sold her soul to the plants for their big money (Grinbaum 2015)?

To clarify this process, I decided to focus on the dialectics of the third party, as described in the book *Critique of Dialectical Reason* (Sartre 2004). Jean-Paul Sartre developed the concept of reciprocity and the dialectics of the third party in order to explain the manner in which groups are formed and organizations established (Gordon 1985; Matarrasse 2001). According to this approach, the third party is responsible for creating the objective reality in which the parties can discover the reciprocity between them as a situation, that is, as a social entity that has existence in a specific time and space. Familiarity with Sartre's ideas led me to conclude that thinking about a mediation process like this is not possible without understanding the dialectical process that Givon, herself, had to undergo. In other words, I argue that Givon was not invited from *the outside* as an objective and impartial mediator, but rather the opposite. One cannot think about the existence of such a complicated mediation process without understanding the importance of a third party, a person or group of people who are willing to think out the things *from the inside*, and to undergo the dialectical process themselves.

The discussion over a dominant woman who contributed so much to nature and the environment in Israel may clarify a thing or two on the connection between women and nature. Moreover, the discussion from the perspective of a Sartre's third party may also contribute, as Sartre's claim that Nature, or the indifferent material, often serve as a third party in dialogues that take place between individuals and between groups. Ecofeminists, such as Vandana Shiva (1991, 1993, 1997), indicate that women's bodies and even women themselves are often perceived as an indifferent material designed to endure the hardship of life. But these ecofeminists usually do not relate their findings to Sartre's idea of the third party and therefore miss the passive power of such a role.

In this context I will not pretend to explain all the implications of such perspective but rather to illuminate ways of thinking that can help the development of human and sensitive environmentalism. I will focus on Sartre's concept of 'reciprocity,' a key concept for understanding the *triadic* reciprocity of the third party. Sartre pointed to 'alienated reciprocity' as a manner of existence that has characterized man throughout most of human history. The manner of existence that characterizes many in modern-industrial times he terms serial or 'sequential alienated reciprocity' (Sartre 2004; Gordon 1985; Barnes 1992). Through examples from the life of Givon, and while concentrating on the events that led up to the mediation process, I show how Givon developed an understanding of alienated reciprocity as characteristic of the environmental movement itself and not only as a characteristic of the owners of the chemical plants or state officials. Following that, I explain the concept of the third party in connection to nature and the environment, and I describe the events leading to the mediation process while emphasizing moments of triadic reciprocity and how they influenced the perception of reality. Finally, a further clarification of the concept of the third party will be made from Givon's point of view following a serious illness she suffered in this period. Two texts written by Bilha Givon herself will serve as a source of examples. Both were submitted as part of her graduate studies in Conflict Resolution (2004, 2010). In addition, I relied on interviews given to the press and an interview made with Asher Grinbaum (2015), one of the senior industrialists of ICL (Israel Chemicals Ltd), and a partner in leading the mediation process from its initial stages. To complete the picture, I conducted an in-depth interview with Givon (2015).

From alienated reciprocity to a reciprocity of dialogue

According to Sartre, reciprocity is the simplest relation between humans, between concrete individuals, each with his or her specific project. However, Sartre's concept of reciprocity is very different from the conventional concept, which sees reciprocity as identification or empathy toward the other (Gordon 1985). In the following quote, Sartre described an abstract pattern of reciprocity that can take place in thousands of concrete examples:

> Two men are performing a certain task together. Each adapts his behavior to that of the other, each approaches or withdraws according to the requirements of the moment, each makes his body into the other's instrument to the extent that he makes the other into his, each anticipates the other's movement in his body, and integrates it into his own movement as a transcended means; and in this way each of them acts in such a way as to become integrated as a means into the other's movement. But this intimate relation in its reality is the negation of Unity.
>
> (Sartre 2004, 114)

We learn from this quote that reciprocity, when you strip it of all moral aspects, is the reference with which a person refers to himself and also to others as a means. Beyond the historical fact that it goes against Kant's imperative to treat each person as an end rather than a means, it is important to clarify that Sartre is not trying to develop a code of ethics or to turn us into people with greater awareness and patience, but quite the opposite: he wants to show the situation in the world as it really is and not as we would like it to be. He seeks to deal with the least pleasant moments, when individuals create a hell through their own praxis: through tools and instruments they reify and objectify themselves, referring to every creature, whether human or not, as present-to-hand. The examples are many and they are also important to the environmental movement.

Sartre tells the story of a Chinese peasant who cuts up slivers of wood and thereby contributes to the disappearance of the forest, the erosion of good soil and the clogging of rivers, that later cause dangerous flood. In another example, he describes a woman who works on instruments that produce cosmetic products, while she dispels boredom by dreaming about sexual adventures. This peasant and this woman serve as an example of how people turn their own bodies into instruments of the other while turning others' bodies into instruments of their own. These serve as examples of how an individual can create his own deprivation: the peasant produces floods that will devastate the land, and the woman produces the cosmetic products that she herself will not be able to buy. But unlike other Marxist approaches, Sartre is not satisfied with the materialistic aspect of alienation, but expands and explains the various aspects of alienated reciprocity, which at its base is the reason the shortage is created. Thus alienated reciprocity, which as we have mentioned is characteristic of most of the conduct of people around the world, is, according to Sartre, the main reason for the permanent shortage that people feel in the world. Therefore, Sartre concludes, this shortage is not imposed on them in advance, but is gradually formulated as a result of a praxis of shortage that is guided by baseless greed and bad faith.

One can question whether Sartre was not exaggerating a bit when he asserts that for most people reciprocity with others and with the world stems from bad faith, a faith that leads to alienated reciprocity. In any case, the period in which Givon decided to leave the SPNI, to establish the Sustainable Development for the Negev organization and to return to dealing with the chemical plants at Ramat Hovav, was also the period when she began to become aware of the alienated reciprocity of the environmental struggles. Read the words with which she describes this period with industry (2010). At first she describes her successes but later she doubts those successes and her ability to cope with the alienated reciprocity of the developers; finally she questions the alienated reciprocity of environmental organizations and their whole attitude toward nature conservation:

> From 1983 and until 1994, I specialized in and developed the environmental struggles in the Negev. From 1994 to 1999 I dealt with the nationwide struggle. These environmental struggles were without compromise, and to their credit the Ashdod sands, the craters, the Jaffa port and the Tel Aviv seashores all exist today. In the Arava region we prevented construction of the "Voice of America" station and the airport at Ein Evronah. We prevented the paving of a road to Ma'aleh Shacharut and the paving of an additional road to Masada. We prevented establishing the city of Hever along the shore of the Dead Sea. We led the legislating of the Beaches Law and prevented at least ten marinas along Israel's shores and much more.
>
> Yes, I was good at this and achieved many successes, but, when I analysed these successes it suddenly hit me: a few battles were indeed successful and prevented the project, but it was prevented not because of the battle but because the situation changed. . .
>
> My successes were the struggles, and I did not consider this a success. They saw this as a success and I viewed it as a failure. Once a plan that we had succeeded in preventing returned – a plan for building along Israel's shoreline, or a plan that nibbles away at the nature reserve on Mt. Carmel – and they always return after a few years, when such an issue is back on the agenda, it is a failure. It is a sign that perhaps we managed to delay the building, but we did not change people's attitudes. But in the Green Movement it is not seen that way, because people are replaced, because headlines are needed in order to get funding. The Green Movement lives off these battles.
>
> (Givon 2010)

In an interview I conducted with her, Givon was even more decisive concerning the alienated attitude of the Green Movement as a whole and especially of her own role in this framework. Perhaps after attempting for so long to prove the success of the mediation process with the factories, she felt more confident about casting doubt on past and present experiences:

> In these fights, you are on the side that is right, and this is something that gives you strength, particularly when the objective is conservation of nature and the environment. In this fight you know what is best for the public; there are no doubts or questions. This fight puts you in the headlines and you bask in the exposure, especially when media sees you as the "good guys" and the other side as the "bad guys". The other side is the one that harms nature, the landscape and the environment. It destroys out of greed for money while you act without any motives, political, economic, or otherwise. Your hands are "clean".
>
> [In the Green Movement] they missed the part of the change in perceptions. Perceptions don't change. Each year a million people pass through SPNI [in trips, nature activities, free guided tours]. The organization exists for over 60 years . . . that is 60 million people! We should already be drinking mineral water from the streams. . .
>
> In all of the struggles we had for the sake of nature, people did not interest us, not even the registered members of the SPNI. We used them. We thought we knew what was good for them and used them. All of the decisions were ours. We did not concern ourselves with what the [local] residents felt. After all, if the chemical plants were shutdown tomorrow, then we would have a Negev free of toxic materials but the people would have no reason to live in the Negev.
>
> (Givon 2015)

In the interview, Givon dated the turning point to the year 2001. At that time, she returned to deal with the issues of Ramat Hovav, after being absent from the Negev for many years. The immediate reason for her renewed preoccupation with this subject was the irritation caused by the unbearable smell that wafted from the plants' communal evaporation pools of industrial sewers. For many nights, the approximately 200,000 residents of Be'er Sheva had to shut themselves indoors with windows closed, due to the inversion phenomenon that brought the cloud of pollution directly into the city. Attempts to solve the chemical waste problem through induced evaporation—spraying the polluting industrial waste into the atmosphere in order to accelerate the evaporation process—created additional problems, and devastated another region. The construction of a facility for the biological treatment of industrial chemical waste, at a cost of over 55 million shekels, also failed to provide the desired results; the facility stopped operating after two months, and stands today as a white elephant (Lichtman 2002).

Alon Tal, in his book *The Environment in Israel* (2006), argued that the policy of the Ministry for Environmental Protection in those years lacked focus and was devoid of economic and legal instruments that could provide a solution to the variety of problems related to the hazardous materials. For example, the Hazardous Materials Law, passed by the Knesset, is far from modern legislation that is able to meet diverse problems that come with the production of hazardous substances. There is lacking a control system for toxic waste "from the cradle to the grave," and the Ministry for Environmental Protection, instead of adopting standards and technologies that have proven their effectiveness in Europe and the U.S., preferred to treat the symptoms and not to focus on the in-depth problems of production, transportation, burial and combustion of hazardous materials (Tal 2006, 433–434). The plants themselves were helpless. On the one hand, investment in treatment requires large sums of money and did not provide the desired results. On the other hand, enforcement measures were increased and they found themselves fighting in court over excessive emissions of pollutants and accidents that were not reported in time (Lichtman 2002; Waldoks 2008).

In light of this situation, it was not difficult to mobilize the environmental movement, to block the roads leading to the industrial zone, burn tires, and mobilize the media against the chemical plants. Givon was highly regarded because of her organizational and confrontational abilities. Thanks to these skills, she was invited to meet with government ministers and representatives of the industrialists. In the interview, I asked her to focus on the feminine aspects of these encounters. To my surprise, Givon chose to compare the attitude toward women and the attitude toward non-profit organizations of the third sector, even though they are not all run by women:

> At first I felt that they measured me by the size of my breasts. This is always true. You arrive at the committees and they assess you according to exterior form. Afterwards there is the disdain: what do you know, what do you understand? But it is not only as a woman; I have a feeling and this should be further considered, that they treat NGOs with disdain because they have no influence, they are not connected to the government and they don't represent the government. They proposed, several times, that I work at the Ministry for Environmental Protection, and the Ministry for the Interior, but I refused. I wanted to be an outsider; I wanted work from the outside. When you are part a system you have to fall in line with the system.
>
> (Givon 2015)

According to Givon the turning point came in 1999, when Akiva Moses was appointed CEO of Israel Chemicals Ltd. A short time later, Givon received a golden envelope containing an invitation to the Environment Prize Award ceremony, which would be awarded to Moses by the Manufacturers Association of Israel. Moses, out of respect, invited her to the ceremony and Givon came. There, she says, the ice was broken when the mutual respect became tangible and overt by a simple human gesture. But this mutual respect was achieved in the price of a battle she waged against him in the 1990s, when he served as manager of the Zin and Oron plants at Rotem Amfert:

> I knew Akiva Mozes when I conducted the fight against the mining of phosphates in Zin and in the Gov stream. The stream serves as a popular hiking route, therefore I asked the ranger from the Nature Reserves Authority to report to me each time they come to mine phosphates within the area of the nature reserve. The Nature Reserves Authority did not want any problems with them but I was not afraid of them. I went to check if they had permits and I saw a caption on the file at the Ministry of the Interior that they had a permit because they were a "Zionist plant". Imagine, they had no permit whatsoever! I stopped all of his [Akiva Moses] work, and I made them prepare plans for mining phosphates, which was never done until then.
>
> (Givon 2015)

The discussion so far has revealed that Givon's decision to seek another way, a way of dialogue, was made against the backdrop of disappointment that became more and more distinctive of the alienated reciprocity that characterized the environmental battles. When Givon requested an appointment with Moses, she hoped that "perhaps this will lead to a better outcome than a situation [where] everyone fights against everyone" (Givon 2015). This all-out war was also summed up by Moshe Lichtman (2002), a dedicated volunteer at the Sustainable Development for the Negev Organization. The vast experience she accumulated allows her to outline a better vision for the environmental movement:

> We [in the environmental movement] find it difficult to conduct battles over a period of time, so we do it in bursts. Therefore, work must be done on changing perceptions. We don't have fancy lawyers, time or resources. So we cannot run to the courts before we tried other ways. We must also be honest and express the will of the public rather than our own will. Therefore, it is unacceptable for green NGOs to make money out of lawsuits [as one of the green NGOs does today]. That is a move that will be used against us in the future.
>
> (Givon 2015)

To conclude this topic, it can be said that, whether or not Sartre was right in claiming that alienated reciprocity is the mode of existence most common throughout human history, it is important to show both the extent of the alienation and the process of sober awakening that Givon underwent in this context. On the other hand, one can ask whether Givon could reach the mediation process without obtaining what she calls the "reverence," or the feeling of "respect and suspect" (Givon 2015), that was established between her and the industrialists? In other words, could Givon have become the authority, or a dominant third party in the mediation process, without the courage, determination, and cleverness that emerged and developed during those struggles? Below we will get to know Sartre's original idea of the third party, which is not necessarily a peacemaker, a mediator, or a conciliator. As with the concept of reciprocity, it seems that this term actually helps to rethink Nature in a new context. Then, the events leading to the mediation process will be described while emphasizing moments of triadic reciprocity and how they influenced the process of environmental dialogues.

Nature and the dialectic of the third party

In the best-known example Sartre gave for the concept of the third party (Rio 2002), he describes a situation which occurred during his stay on vacation at a hotel. While looking out of the window, he noticed two workers, one of them was a gardener, trimming, hoeing in the garden of the hotel, and the other, a road-mender repairing the road. The two workers were very close but were not aware of each other as a high concrete wall stood between them. Thus, at the same time as Sartre, the holiday maker, totalizes the situation of reciprocity between the two workers, he realizes himself as a partner in a triadic relation and he situates his own relation to the two workers as negative: "I do not belong to their class, I do not know their trades, I would not know how to do what they are doing, and I do not share their worries":

> By this I mean that the reality of the other affects me in the depths of my being to the extent that it is not my reality . . . I see the two people both as objects situated among other objects in the visual field and as prospects of escape, as outflow points of reality . . . Each of the two men is reconceived and located in the perceptual field by my act of comprehension; but with each of them, through the weeding, pruning, or digging hands, or through the measuring, calculating eyes, through the entire body as a lived instrument, I am robbed of an aspect of the real. Their work reveals this to them and in observing their work; I feel it as a lack of being. Thus their negative relation to my own existence constitutes me, at the deepest levels of myself, as definite ignorance, as inadequacy. I sense myself as an intellectual through the limits which they prescribe to my perception.
>
> (Sartre 2004, 101–102)

Gordon (1985) explained that one could argue that in these excerpts Sartre is merely going beyond what he already explained in the section on "The Look" in *Being and Nothingness* (Satre 1984). There he discussed the Other as an object in one's perceptual field; here he is discussing him as a person performing praxis. But this seemingly simple change has great significance: the third party is a key to making intelligible social relations and developments. For instance, if the road-mender and the gardener are ignorant of each other, it is an ignorance which exists through Sartre the 'holiday maker.' He is their objective milieu at the moment of his looking out of the window at them working. Similarly, Gordon concludes, the teacher will be the objective milieu for his two pupils when he calls them to account for acting cruelly toward each other.

Their cruelty becomes a social reality through the teacher's admonitions, through his performing his praxis in educating their character. The teacher, as a dominant third party, can be the person that enables the dialogue between the pupils, but may also be the one that allows only sequential alienated reciprocity, as is often the case in overcrowded classrooms managed by outdated educational methods.

It may be argued that after the laborers or the pupils meet and establish a reciprocal relationship between them, there is no longer a need for a third party. But Sartre insisted that the third party, as the person responsible for the objective reality within which the parties can reach a dialogue to reveal the reciprocity between them as a situation, will continue to accompany all of the reciprocal relations between them. However, this accompaniment will not come *from the outside*, such as a teacher who observes and supervises pupils. The triadic relationship is a relationship that the participants experience *from the inside*, a relationship of *practical immanence*. This relationship is embodied in the language they share, in gestures such as a handshake, and in the tools they will exchange. Language, tools, and even the environment itself can be used as a third party because they are manmade. Therefore, people are able to experience the motivation that led to their creation, internally.

From the above we can understand that a moment at which a tertiary relationship is formed is an existential moment that has occurred with certain people at a specific space and time, a moment after which history appears different. Sartre gave as an example the moment when the Emperor of China boarded a plane to see for himself how the collective praxis of deforestation in China, a praxis that took place over several generations as part of the effort to expand its farming areas at the expense of the nomadic herders of the desert, causes the phenomenon of erosion of soil and flash floods from which all the residents of the area suffer. Until then, every farmer thought he was acting alone and did not understand the wider context of his actions. The new perception that was formed during that fateful flight changed the reciprocity between the Chinese themselves and between them and the nature of their country. In this sense, we can understand why establishing a triadic relation is an accomplishment and a moment of no return. Moments where individuals serve a third party for others are moments incised in memory, moments without which one cannot understand the course of history, whether his own history or the history of mankind.

The mediation process and triadic reciprocity

We have seen above how the understanding of an "alienated reciprocity" influenced Givon's decision to conclude her environmental struggles, to give up on her presence as a Green observer in regional planning and construction committees, and to start looking for ways of dialogue with the industry. I also mentioned that, according to Sartre, the dialectical process of the third party is an internal process, a relationship of practical immanence. In other words, the manner in which a person serves as a third party to other people passes through his own materiality. Indeed, we have seen how Givon became acquainted with key individuals in the Green Movement, in the Ministry for Environmental Protection and in industry. The fact that she has lived in the Negev for decades and worked as an educator for years has turned her into a recognized and well-known personality. Givon frequently gives interviews to the press and to the media, as part of her environmental and fundraising campaigns; I, however, have found no reference to these biographical details, which I knew indirectly from my familiarity with her. I, therefore, held an in-depth interview in which I asked her to focus on how she had to change her lifestyle in order to be able to play a major role in the mediation process.

Despite the importance of Givon's intentions to establish a dialogue with the industry, it appears that the big change that enabled her to go through the mediation process was connected to the establishment of the SDN as an independent organization. Givon was about fifty years old then; she had a respectable position and a steady job in the nature conservation department of the Society for the Protection of Nature in Israel (SPNI). According to her, in that year, 1999, it was not her intention to leave this job in order to embark on an adventure that would also involve her financially:

> The idea was to establish a branch of SPNI in the south. The funds that provided the finance did not agree. They wanted the money [to go to] the organization I established. I consulted with the CEO of SPNI and received his blessing just as I had received the blessings of many other people, but the decision was about me. If I had known then what I know today about managing organizations, raising money, preparing a budget, training people and so on, maybe I would not have gotten into it at all.
>
> (Givon 2015)

One of the first acts of the SDN was to build a think tank. Involving more than sixty experts, academics, and environmental activists, think tanks were created to engage in a variety of subjects: chemical waste, urbanization, regional councils, environmental education and so forth, a process summarized in the booklet titled *Principles for Sustainable Development of the Negev*: "to initiate and support the general population in sustainable way of life based on positive change and development through dialogue, transparency and public awareness preventing and repairing environment degradation" (Givon 2001, 2). As previously mentioned, during those years, the problem of bad smells emitted by the evaporation ponds of Ramat Hovav evolved. In 2001, nineteen factories operated within the Ramat Hovav Industrial Council area, producing about 50% percent of all chemicals produced in Israel. Also operating there was the national site receiving all toxic and dangerous waste from the entire country. Adjacent to this site, an incinerator for burning these hazardous substances was erected. The effects of chemical poisons were already clear in those years; scientists expressed concern about genetic changes, congenital diseases, aborting of fetuses, infertility, malignant and degenerative diseases, damage to water and food reservoirs, and more.

Givon identified the main issues concerning Ramat Hovav as early as the late 1970s, the years when the first plants were built. But now, after years of struggle against those chemical plants, she sought a different direction. As mentioned above, it soon arrived in the form of a golden invitation, and Givon did not miss the opportunity to start afresh with Akiva Moses, CEO of Israel Chemicals Ltd. With the establishment of SDN and the support of more than sixty registered members, she received the blessing of Akiva Moses and Asher Grinbaum, and negotiations to establish the Community Advisory Panels (CAP) began soon thereafter. This is how she describes those negotiations:

> At first, contact was made with the CEO of the Bromine Compound Plant. Together with him, a representative of the industrialists and their lawyer, we met and tried to see how to undertake the process. For a year and a half we formulated a vision and drafted the regulations. There were quarrels about every sentence until we set down the contract. In July 2003, we began the first forum at the Bromine Compound Plant.
>
> The greatest obstacle before the industrialists was: what will happen if the greens go to the media when they discover cases where plants violated the law. "If we reveal everything to you [they asked], then who will guarantee that you don't use it against us?"
>
> I understood the plants' demand and first of all I undertook not to disclose the protocols until they approved them. Secondly, I stated that once a relationship is established, I don't run to the media on every matter. Thirdly, if you were in excess and did nothing in order to prevent a recurrence of the faults, then we will go to the media. These are three unwritten agreements that we did not put in the contract. If you do not handle it and 'kill it when it still small', then I'm going to the media.
>
> (Givon 2015)

These citations exemplify how the media served as a dominant third party in the reciprocity that began to form between the Green Movement and the industry. Givon's vast experience with the media allows her to make its presence felt during the meetings. The media is a present absentee and it seems that it is perfectly clear to those in attendance with what kind of articles they are dealing. Another third party serving as a present absentee in those meetings is a different Green organization, which sought grounds for lawsuits against the Ramat Hovav plants. This alienated approach caused the Makhteshim and Bromine Compounds Plant to fear reports on their shortcomings and thus involved the plants in even longer lawsuits (Waldoks 2008). As part of the dialogue process, Givon had to consolidate the identity of the Sustainable Development in the Negev Organization with a praxis that distinguishes it from other Green organizations. Finally, the contract, too, is a dominant third party, representing the State and the Ministry for Environmental Protection. According to Givon, after a relationship of trust was established, the contract and even the media had, to a great extent, lost their dominance.

In an interview with Givon, I asked her to focus on relations between women and men at the negotiating table. Note Givon's remarks on this issue:

> On their part [the industry] there were only men. They knew I was a fighter and they feared me. Even today they speak mainly to me. They don't speak to other volunteers in the forum. Do you see? It's amazing. That is why I am apprehensive about what will happen when I leave. In my opinion, what the industrialists first saw in front of them was "Bilha the sucker and she is going to falls right into our hands" . . . I have no doubt that the policy of the plants to cooperate on the matter of the negotiation was out of an intention to win some media "silence," but we won an amazing improvement on the matter of the quality of the environment. My deep acquaintance with the field of struggles created a "balance of terror."
>
> (Givon 2015)

After the CAP in the Bromine Compound Plant proved to be efficient for establishing the reciprocity between the Greens and the industrialists, the industrialists began to view Givon as one of them. But then there was a turnaround. When Givon decided to join the respondent with the Ministry of Environmental Protection against the factories, this move was seen by them as a betrayal or as they said: "you are shooting us in the foot" (Givon 2015).

According to the SDN organization website, in 2004, after intense activity by the organization to expose the members of parliament and the government ministers to what was happening at Ramat Hovav, the Minister for Environmental Protection came to meet the protesters. In the interview Givon mentioned this brief meeting as a moment of triumph:

> Tzahi Hanegbi [Minister for Environmental Protection at that time] asked me bluntly, "Will you stop shouting and tell us what you want?" I was prepared for that moment and I had three requirements of which I am proud of to this day: first, separation of the effluents – each factory will treat its own chemical waste; second, a system for monitoring and control that is exposed and open to the public, every citizen will be able to know at any given moment what he is breathing; and third, amendment of the law concerning the Ramat Hovav Industrial Council, so that it will enable having green representatives at the council meetings.
>
> (Givon 2015)

As the saying goes, "success has many fathers." For Givon, the sense of success came from the fact that three months after that brief meeting, the major factories in Ramat Hovav received a message from the Director General of the Ministry, advising them to prepare to separate their effluents; each plant will treat its own chemical waste by the target date set for 2006. The proximity of these events gave her the feeling that even if the idea of separating chemical waste effluents was already known in other countries, her determined stance at that meeting with the minister is what ultimately caused its implementation in Israel. On the other hand, a sense of anxiety might have been accompanied this triumph when she realized how complicated the situation had become. At that time, the plants did not have any possibility, of meeting the target set by the Ministry for Environmental Protection. Following those directive requiring each plant to treat its own chemical waste by 2006, the plants filed a petition with the court against the Ministry for Environmental Protection and against the Ramat Hovav Industrial Council. Givon and her friends decided to side with the Green and SDN organization join the respondents to the petition.

However, a problem remains with the Industrial Council of Ramat Hovav, and also among the defendants. Originally the Industrial Council was affiliated to the industry rather than the Greens but following the petition of the chemical plants they found themselves between the hammer and the anvil. Note the way Givon outlines a new framework for them and thereby facilitates their joining to the Greens:

> When I asked the Minister for Environmental Protection, Tzachi Hanegbi, that every plant will treats its own chemical waste, the Director General of the Industrial Council stood next to me and said, "Bilha, what are you doing? You are shooting us in the foot." Under the Industrial Councils Law [the Industrial Councils Ordinance, Book of Laws 2332, 5772, page 136], one of the main functions of the Industrial Council is to treat the waste from the plants, something that constituted one of the main sources of finance. I told him [the Director General of the Industrial Council] that they will constitute a sort of frontline command of the Ministry for Environmental Protection in Ramat Hovav. Instead of treating [only] sewage they will treat environmental quality problems. It was a change of the manner of thinking. The introduction of new technologies created new roles, and today there is a twelve-man department in the industrial council dealing with matters of environmental quality, and its budget is about one third of the budget of the Council.
>
> (Givon 2015)

When SDN organization joined as a representative of the public to the respondents of the petition, Givon attended the hearings until 2004 when the new Minister for Environmental Protection, Shalom Simhon suggested a process of mediation (Rinat 2005). Prof. Avishay Braverman, president of the Ben Gurion University of the Negev, was elected as mediator. But in practice, another official conducted the process. Under his management, the matter tangled and delayed. The situation seemed to come back to an impasse:

> It turned into a lot of pointless discussion and it was dragged out until 2006. I decided to intervene when ideas that were clearly non-environmental

> started popping up. Since the Ministry for Environmental Protection demanded that not a single drop of liquid exit the plants, the plants claimed that this could not be done; it's impossible. At the end there will be brines that will need a place found for them. An idea was raised for getting rid of the brines by a pipe that would flow to the Mediterranean Sea, or to the Dead Sea, or even to the Zin Stream. I wrote a letter and demanded that decisions be made. In the end, the decision was made by the professional level, representatives of the plants, and representatives of the Ministry for Environmental Protection [who all] sat together, "turned over the entire world" and found an appropriate technology.
>
> (Givon 2015)

If we stick to Sartre's terminology, it is possible, I think, to discern Givon's good faith. Whenever she feels that she has reached an impasse, she is willing to learn, consult, and modify her basic tenets. In addition, Givon has a special talent for surrounding herself with good advisers, like those sixty experts, academics, and environmental activists, whose advice contributed to the shaping of the image of SDN at its inception. Even today, there are 150 registered volunteers, most of them retired personnel of chemical plants, participating in twelve public forums in industry. These advisers, friends, and volunteers believe in her and empower her as she undertakes her courageous decisions. While critics argue that she is a soloist, working alone and making decisions on her own, justifiably arguing that she has failed to train another person to fill her place in the future, they do not deny the central role she has played, since the year 2001, in the establishment and management of eleven public forums in industry. Asher Grinbaum, Deputy CEO of I.C.L., testified on this matter in an interview he recently gave to Refaela Babish:

> All that is seen at Ramat Hovav is the fruits of this partnership with the residents and especially the dialogue in the Community Advisory Panels. These things are fine, but what is needed is the proof, the investment . . . we were not able to duplicate it in other regions. Because, at the end of the day, what is needed is this personal connection . . . [people] like Bilha, that really believes that there is no deception. This mediation can only work when trust is created. If one day, members of the [CAP] discover they have been deceived, that things were hidden from them, that it is all "green wash," at that same moment everything will fall apart. In other regions, in Ashdod and in Haifa, there weren't any leadership figures who took the reins out of a willingness to completely strip down, and I blame the industry for this. I told them, you are suspicious all the time. Please, talk! Although every case is different.
>
> (Grinbaum 2015)

According to Grinbaum, the huge investment in technology designed to prevent chemical contamination was the proof of plant's truthfulness of intent. As a member in one of those public forums, I can attest that factory managers are very proud to show and explain to SDN volunteers about this advanced technology. It is therefore possible to say that this new technology enables the triadic reciprocity established by members of the CAP with factory managers. Oddly enough, Grinbaum's rich experience turned him quite skeptical regarding the possibility that such process will take place in other areas of Israel. He points to Givon as "a one-time leading figure who made the factories open up and talk." This issue increases the mystery, because as far as I know nothing in Givon's formal training prepared her to serve as a mediator in such a process. Givon is not an engineer; she is not a chemist, and certainly not a legal expert. At that time, she did not even hold a mediator diploma. Why then she decides that if environmental national codes are ignorant of industrial needs, it is an ignorant that in some way passes through her? Or perhaps, her ignorance of chemicals and engineering was actually an advantage because from the standpoint of a layman or the public, she could serve as an objective milieu for those experts, calling them to account for acting recklessly, wastefully, and irresponsibly?

Following is Givon's description of her personal life at that period. It seems that her personal experience gives credence to Sartre's claim that the process of the third party is an internal process, a relationship of practical immanence. We mentioned that according to Sartre 'the holiday maker' is not just an observer intervening in other people's life. A moment of triadic reciprocity is experienced by the third party as "a lack of being" as a negative relation to his own existence, as a relation that constitutes him at the deepest level of himself. Thus the holiday maker crystallizes his identity as an intellectual who may assist the workers in fighting for their rights. The final chapter is devoted to this idea in an attempt to trace the process that enabled Givon to 'betray' both the Greens and the industrialists in order to truly represent the public.

Givon's practical immanence as the third party

In 2001, at the height of the process of setting up the CAP in the Bromine Compound Plant, Givon contracted colon cancer and doctors predicted she had three months to live. At the same time, the organization faced a financial crisis because the funds that provided the finance suddenly stopped their funding. The organization was in debt for 270,000 shekels. At that time, Givon's eldest daughter also got divorced and moved back into her parents' home. The aid to her daughter included caring for her infant son as she was bedridden due to a high-risk pregnancy. Givon describes this period as a crucial stage in her life where she had to deal with many contradictions. See how she describes the contradictions in the different requirements demanded from her as a wife, as a mother, and as the manager of an organization, and through dialogue with her husband, Yehudah, she negates the trivial and concentrates on what is essential:

> I thought my life was about to end and what kept me going was the attempt to get the organization out of the situation it was in, to have it

> financially balanced, and to leave. I loaned money from home and kept silence about the financial state of the organization. Some industrialists earned in one week the entire amount I owed, but I disclosed nothing. I did not want them to think that we needed their money. A relationship was established based on mutual respect or a sort of equality, and it was important to keep it so. I was receiving a course of chemotherapy that was making me weak, and going to give lectures in order to earn another 500 shekels for the organization. I refused to give up because the feeling of failure burns. Together with all this, there was a "rebellion" within the management of the organization, three of the nine members of the executive committee did not agree with the process of negotiations with the plants. They wanted to fight! The other six members told them that if Bilha leaves, then there is no organization, but if they leave, then the organization will overcome this, and so it was. Three members of the executive committee left and maybe even contributed to defaming the organization among green organizations and in the media.
>
> . . . It completely changed the perceptions, all the matter of the illness, and all these blows . . . first of all, my attitude was not to listen the vilification of others. It simply doesn't matter, because if you believe that your way is the right one, then go your way. Life is fluid and can end at any moment. You can't know what will happen to you tomorrow. It took me a lot of years to build my self-confidence . . . I had no problem of financial security because Yehudah always earned three times more than me. He was the chief breadwinner. But still, the feeling of partnership is important. You need to justify it. But you can justify it even with a third or a quarter . . . but this was a joint decision with Yehudah. I did not do it without his knowledge. Yehudah had faith in me that I would succeed in raising the money.
>
> . . . I never saw money as something. When I see something I want I don't even check how much it cost . . . In the SPNI I never had to deal with anything to do with money. But in the organization I have to deal with manpower, with budgets, with raising funds. This is one reason why I would not establish an organization today. It limits you very much. An organization is the idea but it is difficult to realize the idea if you are busy with other things such as raising money, manpower, bookkeeping. Also, if I want the organization to survive, I need to do activities that are not always in line with the real objectives of the organization. As an organization, there are limitations.
>
> (Givon 2015)

As I mentioned, I had to hold an in-depth interview with Givon in order to know these hidden details behind the scenes of the mediation process. I was particularly amazed to hear how, even on the verge of death, she adhered to her life's mission and refused to give up. She courageously faced death refusing any thoughts of despair and resentment. In this context, it seems that Sartre would not agree with Givon's argument that the establishment of an organization makes it difficult to realize the idea. It appears that the opposite is true. The organization as an institution that exists beyond the existence of this or that person changes the situation in many ways. For example, we saw how it enabled Givon to establish an identity of "Greens" who are capable of dialogue and who are capable of acting out of a long-term vision that requires patience and perseverance. In retrospect, it seems that Givon's decision to completely give up the trappings of honor and money in order to determinedly maintain what had been achieved paid off when, in 2003–2004, other ten factories joined the movement of CAP.

Ultimately, Givon's standing by the Ministry for Environmental Protection and against the plants helped create the status quo that exists to this day. The change that was generated at Ramat Hovav did not happen all at once and certainly was not brought about by one woman. But, as someone who has accompanied Givon for many years, I can attest that the change in consciousness that occurred in the first decade of the 21st century was enormous. The main motto, as it is expressed by the managers of the plants, is the importance of safety and environmental protection (ecology in their language). This is actually the principal change, because in the past, production and making profits were the top priorities. From meeting to meeting new ways to embed the values of safety and of ecology can be seen. Seminars are conducted, rules are written, automatic monitoring systems are purchased and emphasis is placed on physical conditions that will encourage employees to see the immediate connection between maintaining safety at work and environmental protection.

Conclusion

Bilha Givon is one of quite a few women running and managing a major environmental NGO in Israel. Her devotion and dedication over many years to environmental issues have made her a well-known figure but no less controversial. Asher Grinbaum's claim that Givon is "a one-time leading figure who made the factories open up and talk" can raise a few eyebrows: what can we learn from a one-time individual? Within this chapter I wanted to expose the one-time individuality of Givon. How does she reconcile between nature and public concern? How does she resolve her femininity with her determined confrontations with the industry, which is predominantly male? The ability of women to serve as the third party and thus motivate complex processes often stems from obstacles and constraints that they face. For example, the willingness to give up a respectable income allowed Givon to be on the 'right side' from which it is easier to rebuke and admonish factory managers for their inability to decide and implement environmental national codes. On the other hand, the establishment of SDN organization appeared as a wise decision because the NGO itself serves as a third party for complex processes. Over the years, several SDN employees who started without any environmental training moved and became valued employees in the Ministry of Environmental Protection and the

City Council of Beersheva. Even for me, as one who worked with Givon thirty years ago and currently is a board member of SDN, Givon is a 'source of values,' or as Simone de Beauvoir said:

> For existentialism, it is not impersonal universal man who is the source of values, but the plurality of concrete, particular men projecting themselves toward their ends on the basis of situations whose particularity is as radical and as irreducible as subjectivity itself.
>
> (2006, 17)

Note

1 Following the success of the mediation process, it was decided to change the name of the industrial zone to Neot Hovav—Eco Industrial Park in the Negev.

References

Barnes, E. H. 1992. "Sartre's Ontology: The Revealing and Making of Being." In *The Cambridge Companion to Sartre*, ed. C. Howells. Cambridge: Cambridge University Press.

De Beauvoir, S. 2006. *The Ethics of Ambiguity*, trans. Bernard Frechtman. New York: Open Road Media.

Givon, B. 2001. *Principles for Sustainable Development of the Negev*. Omer: SDN Organization.

——. 2004. "Community Advisory Panels in Israel." www.negev.org.il/index.php?m=text&t=&la=&lib=367

——. 2010. *My Notebook of Reconciliation*. Submitted as part of MA studies, Ben-Gurion University of the Negev.

——. 2015. Interview with S. Tamari.

Gordon, H. 1985. "Dialectical Reason and Education: Sartre Fused Group." *Educational Theory* 35(1): 43–56.

Grinbaum, A. 2015. Interview with R. Babish.

Lichtman, M. 2002. "Ramat Hovav: Talking and Talking While Poison Seeps." *Globes*, January 20.

Matarrasse, Ć .2001. "Solidarity and Fear: Hegel and Sartre on the Mediation of Reciprocity." *Philosophy Today* 1(45): 43–55.

Rinat, T. 2005. "Ramat Hovav Manufacturers: Will Transport Factories Abroad." *Haaretz*, March 13. http://news.walla.co.il/item/684305

Rio, K. 2000. "The Sorcerer as an Absented Third Person," *Social Analysis* 3(46): 129.

Sartre, Jean-Paul. 1984. *Being and Nothingness*, trans. H. Barnes. New York: Simon & Schuster.

——. 2004; orig. pub. 1960. *Critique of Dialectical Reason, Vol. 1*, trans. A. Sheridan-Smith. New York and London: Verso.

Shiva, Vandana. 1991. *Ecology and the Politics of Survival: Conflicts Over Natural Resources in India*. Thousand Oaks. Sage.

——. 1993. *Monocultures of the Mind: Biodiversity, Biotechnology and Agriculture*. New Delhi: Zed Press.

——. 1997. *Biopiracy: The Plunder of Nature and Knowledge*. Cambridge, MA: South End Press.

Tal. A. 2006. *The Environment in Israel: Resources, Crises, Campaigns and Policy*. Tel Aviv: Hkibbutz Hameuchad.

Waldoks, E. Z. 2008. "Poison Cloud from Chemical Plant Could've Been Prevented." *The Jerusalem Post*, April 6.

9 "Yo soy mujer" ¿yo soy ecologista? Feminist and ecological consciousness at the Women's Intercultural Center[1]

Christina Holmes

At the Women's Intercultural Center (WInC) near New Mexico's border with Mexico, Sofia sings "Yo soy mujer. Yo soy mujer en busca de igualdad." The song served as an opening ritual to meetings held at the Center and the lines translate to "I am a woman. I am a woman in search of equality," reflecting the development of feminist consciousness among participants. By Sofia's account, the song inspired her, built up her self-esteem, and opened unexpected doors for her; across interviews with staff and participants, women spoke of being transformed by their time at WInC as they came to consciousness regarding the oppressions that Latina immigrants in the United States face.

WInC is a non-profit organization serving women in southern New Mexico and west Texas. Its mission is to "provide a place for women to learn and work together to develop their social, spiritual, economic and political potential" ("Mission" n.d.). WInC offers classes for economic self-sufficiency such as the Small Business Academy; classes for personal empowerment such as painting, sewing, ESL, and citizenship; and workshops on nutrition, violence against women, and feminism. Programs are aimed at skill building and consciousness-raising, but at the heart of programming is an effort to challenge the social isolation rampant in the community. Many WInC participants lived in Mexico before moving to Anthony, and thus left friends and family behind to find themselves isolated in a rural new home with few English language skills. A primary drive behind the creation of the Center was to address these concerns and to facilitate a greater sense of belonging among the women of Anthony. Josie explains, "I live out in the country and my husband is working. I was alone most of the time and I was very depressed . . . I tell everyone, if you come here, you won't be depressed. It's the medication that you need" ("Where It Starts" n.d.).

While not embedded in the mission statement, many participants also develop an ecological consciousness as they begin to theorize issues of gender, race, class, and nation in their lives. Sofia, who spoke joyously of the song above, explains, "Another thing that I learned here: to love nature, to give praise through meditation because we don't appreciate what we have. If you don't appreciate something, you're not going to take care of it. So you have to know about it in order to take care of it." Sofia was present when participants built the Center's main building out of recycled tires and earth. Deeply insulated, it uses little energy to heat or cool itself. The building also features rain barrels, solar panels, recycle bins, and a thrift shop for the recirculation of clothes, shoes, and furniture. Recent environmental programming includes classes on organic gardening, recycling, and water conservation. In practice, WInC works at the intersection of feminism and environmental justice; though the ecological elements are easily overlooked, it offers an important case study to understand the relationship between environmental activism and feminist concerns among some of the most vulnerable populations in the U.S., including underserved women in a heavily policed border region. I argue that WInC mobilizes a borderlands environmental praxis with activities that stretch the self into ecological relationships with human, nature, and spirit *others*; drawing on Gloria Anzaldúa's work (1999), I named this praxis "borderlands ecofeminism" because borderlands theory may be understood to refer not only to the symbolic and material conditions of life on the border, but also to displacement from ecological belonging and strivings to re-root. As such, while environmental practices such as recycling and organic gardening are important, an informal, subtle form of ecological consciousness that emerges there, one that strives to form new connections to place and community, may have the most to teach us about intersectional feminist environmentalism; in this, borderlands ecofeminism also adds concreteness to some of the more theoretical ecofeminist celebrations of the interconnectedness of life.

Feminist environmentalism in the borderlands: ecological consciousness

WInC prioritizes consciousness-raising where 'consciousness' underscores the importance of reflection on our social lives and exploration of their political implications—an idea that derives from the feminist practice of consciousness-raising (i.e., making the personal political) and from Paulo Freire's liberation pedagogy (1970). To raise *ecological* consciousness is to make the ecological political. First, ecological consciousness requires understanding of the place facts for a given area or bioregion, including knowledge of vegetation, animal species, landscape features such as the watershed, and an understanding of how resources (e.g., water, energy, food) are managed in the region (Thayer 2003; Tweit 1995). Bioregional awareness also accounts for changes to the land and its inhabitants as a result of human interventions (Dodge 1990). It necessitates systems thinking to see how Anthony, once known for its fertile land and occasional flooding from the Rio Grande, is now home to a number of water-intensive alfalfa and dairy farms trying to survive (and contributing to) the desertification of the Mesilla Valley. Drought impacts the agriculture-based economy and increases food insecurity among a population that already experiences poverty. An ecofeminist-informed bioregional awareness would note that women and Hispanic-headed households experience food insecurity at the highest rates in the nation (USDA 2014).

In addition to establishing a sense of place, environmental scholars reflect on the kinds of subjectivity that facilitate ecological consciousness. On an intellectual level, individuals should begin to develop awareness of their place in webs of connection to activities such as goods and energy production, housing sprawl, meat consumption, environmental activism. But ecological consciousness also relies on an *affective* ability to embed ourselves in a place through what Marla Morris calls "dwelling-in-the-world" (2002, 572). To "dwell" is to develop bioregional knowledge and relationships to a place and its inhabitants. Yet playing on the other meaning of "dwell," the practice also incites reflection on our thoughts and emotions, allowing us to understand the effects of place on our bodies and behavior. Dwelling is an embodied phenomenon; perception and emotions can alienate or facilitate emplacement. In sum, ecological consciousness includes both intellectual understandings of bioregionality and systems awareness *and* an emotional, perceptive, and materially embodied connection to human and nature others. Raised ecological consciousness also implies a shift from understanding to action. Although WInC does not have an explicitly environmental mission, the way it creates opportunities to develop ecological consciousness while grounding actions in a Chicana feminist framework of decolonial consciousness-raising provides a model for recognizing and valuing intersectional ecofeminism.

Gendering the ecological self: stretching toward intersubjective embodiment

Several practices at WInC encourage women to redefine their bodies in ways that are not only empowering from a decolonial feminist perspective, but which also demonstrate emplacement—an awareness of "dwelling-in-the-world." Interviews reveal three facets of Center life that challenge the individual to reflect on her connectedness to others and to her physical, material place within the Center, including the cultivation of strong emotionality, bodily awareness, and creativity. On this point, I begin with the most frequently recurring theme: many interviewees spoke of the Center's positivity. This starts with the fact that participants felt safe and supported because they were surrounded by women, but sociality is also encouraged through classes and parties that develop a sense of belonging among the participants. Although the Center has designated spaces for social gatherings and for popular classes such as painting, Reiki, Zumba, sewing, and ESL, there is good flow among the spaces and a central courtyard that connects them; there are always people moving through the Center who are talking and enjoying themselves. Sofia recalls her first experience there: unable to pay a driving ticket, she was court-ordered to work community service hours at WInC; she approached the Center with trepidation. She narrates, "I came in and there was a group of women in that little building talking and laughing. It seemed to me that they were enjoying themselves . . . I felt like it was a place that I wanted to be. I didn't feel afraid or nervous anymore . . . By the time I walked a few steps that first day I was saying, 'I want to be here! I want to be part of this!'"

There are affective and political aspects in efforts to belong; belonging is not just a motion toward embracing others (people, places, non-human others), but can also effect a change in embodiment—an idea that sutures place, affect, and embodiment. Aimee Carillo Rowe writes, "Belonging is that movement in the direction of the other body: bodies in motion, encountering their own transition, their potential to vary" (2005, 27). Through affective ties to others we rethink ourselves together into 'we-ness' and the very borders of our bodies become less firm; we know ourselves through our relations through others. For Sofia, the laughter and positive sense of community caused a longing that resulted in a shift in her identity; she developed a new sense of belonging both to the Center and to her new homeland—a feeling that has consolidated through her long-term commitment to WInC.

In their interviews, many staff and participants also expressed a heightened level of body awareness. First, interviewees spoke of positive changes that overcome the body when one arrives. Those who walk through the door notice the moderate temperature that contrasts with the more extreme temperatures of the desert. During her time as receptionist, Anna saw this: "When you come into the building, you see that it's cool when it needs to be cool and warm when it needs to be warm. And even that affects the way people feel when they go home because a lot of the people don't have central air so they dread going home in the summer." Others claim that making friends and developing hobbies led to a marked improvement in their health. For example, in honing her painting skills, Josie saw her depression and fibromyalgia improve—as her quote from earlier suggests, attending classes at WInC is all the medicine you need! Sandra, spoke to me of her poor eating habits and exhaustion from the stresses of being a single mother. She was enthusiastic about the opportunity to take nutrition and Zumba classes to improve how she felt. In each case, the body is reworked as it moves through WInC and, as Anna's comments suggest, that changed embodiment bleeds into other relationships (to people, places), such as in the home upon return from the Center.

For staff, the female body is a site of potential for physical and emotional health and for relationship-building, but they also recognize the assault on women's bodies from domestic violence, sickness, and hunger as gendered, raced, and classed acts of violence. Elena, a staff member and former participant, elaborates, "Basic human rights-building . . . I think that a lot of immigrants are not empowered. They don't feel confident enough to say, 'You know what, I deserve this' . . . We have citizenship classes that are very important. This office also used to have sexual assault counseling and resources." Because staff understand women's health as a multifaceted and socially constructed arena of concern, their strategies for healing are diverse. From Reiki and labyrinth walking, organic gardening and painting, to sexual assault awareness, and the promotion of citizenship and human rights agendas, they demonstrate systems awareness of the social and physical components of women's health in their programming. Each of these healing strategies opens up different relationalities with others, emphasizing positive connections between people and reworking negative ones such as violence between intimate partners. Importantly, these efforts contribute to ecofeminist theories of embodiment because they highlight women's bodies not as 'natural' but contested social terrains—bodies that reveal and rework how power inscribes and subjugates. Acknowledging this gives women agency to create new scripts and to change their environments.

In addition to emotionality and bodily awareness, WInC values creativity as a means to rework relationships. Carter, the director, explains: "Every woman has something special to share. Women have been known to have incredible talent and creativity so the moment someone discovers a talent, then a class emerges out of that." For example, Rosario came to WInC depressed after the death of her son—she was looking for community. She enrolled in painting classes and eventually began to sell her work; now she teaches her own art classes. For Rosario, the Center "is a support network for me. In being here, it is not only an economic support but it elevates my morale and I feel really happy particularly in sharing my knowledge" (Carter 2009). Classes on painting, sewing, and quilting offer women the chance to create themselves through new relationships to each other, to the materials they work with, and to the world around them that is reimagined in their works. At WInC, these affective connections not only rework bodily boundaries and human relationships, but they also facilitate emplacement through developing connections with the surrounding environment. In the next section, I consider practices that are explicitly environmental as well as aspects of WInC that may be less so, but which raise ecological consciousness nonetheless. While the latter practices (such as building space consciousness) may seem less obvious than acts like gardening, they exemplify a transformation of the autonomous self into one advocated by ecofeminists such as Ynestra King (1983) and Val Plumwood (1986) in which ties to human and nature others remain central to one's subjectivity.

En-naturing embodiment to build ecological consciousness

WInC integrates ecological awareness into varied daily activities. During parties, participants are reminded that plates and cups should be recycled. When participants asked for nutrition and cooking classes, the idea to plant an organic garden sprung up, and participants created a play to introduce the community to the health and flavor benefits of organic produce. Emphasizing the role many women play as household money managers, a "Go Green, Save Green" workshop repurposed household objects (e.g., turning plastic bottles into decorative flowers). As Carter states, WInC considers "possible activities that reduce our imprint on the environment." The Center's construction from tires rescued from landfills and the desert shows this. The tires are free, and their insulation keeps utility costs down while reducing natural resource use—a reflection on the place facts of the region mobilized to enhance social and ecological health among the human and non-human residents.

Like everything, construction was carried out as an educational activity. Participants researched methods of "green construction," planned out the space, and learned the skills necessary to build the Center. The gallery, which is the primary entertaining space, features a partially finished segment of the wall that exposes the building materials, serving as a visual reminder of the building process. The Center also boasts solar panels, rainwater collection tanks, recycling and compost bins, and an organic garden used by the gardening and nutrition classes. Education activities raise consciousness about the social and ecological benefits of the conservation of resources, the importance of local organically grown food, and of limiting waste and consumption. In one nutrition class I observed a discussion of the expense of fresh, organic food at the grocery store, the relative poverty of the community, and women's responsibilities for buying food and caring for their family and community. Questions arose: What are the healthiest options if you can't afford organic or fresh produce? Are frozen and canned vegetables still good choices? Why is it difficult to buy fresh vegetables when large farms surround the town? This exchange reflects the growing systems awareness that emerges from needs in the women's lives. It also highlights how raised ecological consciousness involves an understanding of the links between class, race, gender, and environmental health.

As a scholar of ecofeminism, I was curious what role (if any) gender played in WInC's environmentalism. Elena, a longtime staff member, thoughtfully replied:

> There has to be a relation. Maybe it has to do with [the fact that] most of our classes we offer to empower women; I think that learning that you could have your own garden and you could save money, you're healthier, which means that if you're healthy, you have a mind to learn, you're healthy to work, you're healthy to go to school and you're healthy enough to be a parent . . . I think it's about empowerment in many ways.

Laura, participant-turned-staff member concurs, "You see the ladies come and plant things because to them, that's helping the earth and they take that home with them. With the knowledge of how to recycle, you're going to go home and find that you don't want to throw that bottle in the trash." This environmental education sparks raised consciousness that participants bring with them into the home and other spaces in their lives. Further, environmental education at the Center isn't merely focused on conservation, but is oriented toward the care of others. This is expressed in the care of nature others—that interviewee went on to say that if we can save a bird by recycling plastics, we have a responsibility to do so. It is also expressed as the care of human others—feeding yourself and your family well, being "healthy enough to be a parent," sharing your knowledge about recycling or water collection with others, building community through sharing a garden. WInC enables women to make the most of their socially assigned roles as networkers, a strategy that empowers women to develop and facilitate more equitable and just relationships within their community and between their human community and the natural environment; however, no staff or participants gestured toward a 'natural' or biological link between women and nature. These links were deployed through daily educational activities to empower women and teach them to change their surroundings.

In addition to gathering place facts and building systems awareness, the development of spatial and place consciousness stretches the self into body/landscape subjectivities (Davies 2000). Space consciousness is ever present at WInC. First, the Center's proximity to a church facilitates a spiritual awareness despite the departure of the Sisters of Mercy several years ago. Second, WInC's location between two large metropolitan areas and near the U.S.-Mexico border contributes to its success and made it a hub for people crossing through. This is nicely illustrated by a participant whose quilting square depicts women with linked arms as they march forward. It reads, "across the desert came a multitude of women"—a phrase that, in the strategic use of the word "multitude," conveys solidarity among a large number of women and references the diversity or multiplicity among them. These examples show how WInC took advantage of its geographic location to bring people together and strengthen community ties, but other examples of space consciousness reveal an effort to deepen not just communal ties, but move toward intersubjective associations that appear to be so necessary for the building of socially and ecologically conscious community.

The gallery is the Center's most emotionally charged space. It is the gathering place for events and features a labyrinth that was researched, designed, and tiled by the participants. The gallery also features an internal window that exposes the tire construction and it houses participants' artwork, including a quilt that stitches together messages of gratitude to WInC. Several interviewees believed that there was something positive and peaceful about the gallery despite its nearness to the reception area that occasionally becomes congested and hectic. There may be two reasons for this. First, as a social geographer, Amanda Bingley (2003) studied boundaries around the self that are negotiated through past memories and current landscape interactions. Participants were interviewed about their relationships with familiar places while engaging in both creative activities and a more structured interview setting. Her interviewees used more vivid language to depict self-landscape relations during interviews that involved the creative and sensory interview experiences (in her case, sand play). These findings mirror my own: creative practices and behaviors that prod us to rethink our relations to others and to the landscape produced a change in the embodiment of the subjects—interviewees became more animated, more likely to speak of themselves in relation to particular places and spaces, more likely to see *relationships* in general as constitutive of their subjectivity. Displaying art, exposed construction efforts, and a labyrinth, the gallery is a space marked by women's creativity, prayer, and reflection; as such, it brought out emotional reactions among WInC participants.

Second, while the social and creative purposes for which it is used contribute to its calm, positive feeling, there is also a piece of folklore that marks the Center and that area in particular. The director describes the construction process:

> Whenever they would get really stressed through the construction of the building . . . they had this ceremony. They would get a couple of pens out and they would start thinking and really try to connect with the idea behind the Center. On little pieces of paper they would write their dreams and wishes for themselves and for the Center. Then they would dump all the little pieces of paper within the dirt and within the tires . . . I always laugh when somebody tells me that as soon as you come in, you have this sense of peace even if you're having problems at home. I really think it has to do with all the beautiful well wishes that are located in the different sections of the Center because there's so much positivity that was given to it and that's how they were able to connect to the vision and the mission of the Center and they were able to refocus.

In this ritualized enactment of the mission, participants forge a connection to each other as well as to future generations of staff and participants through the sharing of hopes, dreams, and contagious positivity. Through affective communion, the women energize themselves in the wake of their physically and emotionally exhausting labor and they do so by focusing on their shared strength and shared futures. It is a conscious performance of intersubjective relationality wherein women redraw their ties with each other and, in putting parts of themselves (i.e., their dreams and wishes) into the soil and tires, they become co-extensive with the Center and more deeply invested in its success. The reiteration of this story to new participants bolsters the effect of space consciousness on the women and embeds them in their environment; they refigure their sense of self as contiguous with the human and nature others that comprise their environment both in the present and in the past as a result. Capitalizing on this knowledge, Carter hopes to introduce additional spaces that draw community together, what she calls "a gathering space that is really open to all." Modeled off el Zócalo in Mexico City, she envisions a large concrete patio that could host dances, farmers' and craft markets, and forums on community development. Carter explains, "That attracts me, that something so simple as an area of cement would provide so much to a community."

These examples show WInC's broader sense of the environment that includes not just the natural world, but also an individual's relationship to natural and built environments; this expands the narrow ideas of ecological belonging that can be found in some environmentalist and ecofeminist work. The gallery and other communal spaces facilitate intersubjective selves that are oriented toward community building and education for consciousness-raising, but they are also working toward more thoroughgoing connections between humans, nature, and the built world that surrounds them. As the interviews show, the staff and participants develop identities that are directly linked to these relationships and, maintained through the affective ties and sense of responsibility for social and ecological justice, these identities are solidified through constant performance of WInC's mission. This is visible in the examples above from the ritualized construction of the Center to the ritualized daily welcomes to the constant reminders about the importance of recycling.

Embracing the spirit of ecological consciousness

We saw the Center's explicit and institutionalized commitment to environmentalism alongside more subtle ecological work (e.g., practices of dwelling); the two move in tandem toward the development of a borderlands ecofeminist subjectivity. Similarly, there is a more overt spiritual presence that coexists with a subtler and more diffuse spirituality—interviews show that the narrative of spiritualty emerging at the Center is one that stretches human ties into the nature and spirit realm, facilitating new body/nature/spirit intersubjectivities. Overtly, the Sisters of Mercy founded the Center and its location across from a church links it with a religious community in the area. In fact, some continue to confuse WInC for a Catholic community center despite its non-denominational status. A multicultural approach to spirituality can be seen in the different traditions celebrated at the Center, including Reiki and meditation, healing arts associated with Eastern practices such as Buddhism; the labyrinth, which can be found in cultures around the world; and prayer, which interviewees described as a reflection and appreciation of life rather than a narrowly Catholic focus on sin and forgiveness. Perhaps, because of the Center's efforts to be spiritually open without imposing any particular belief system on staff and participants, there was no single narrative about religion in place to shape interviewees' responses. One person characterized the spirituality cited in the mission as less about religion and more about "the golden rule"—doing unto others what you would have them do to you. Another spoke of it as a feeling that overcomes you: "If you don't want to go to church, it's okay . . . It's not so much a 'doing'; it's not a verb, it's a *feeling*. It's something like a feeling of calm; you have peace" (emphasis in original). As seen earlier with respect to issues of intersubjective embodiment and human–nature relationality, mobilizing affect seems key to create intersubjective identities—especially in prayer and group reflection. One of the younger participants explicitly named the relationship between affect, prayer, and their ability to build strong ties to others: "I think it [group prayer] is important because you get to understand what they're feeling and the prayers connect you more to everyone else. For me, when I pray with my family, it connects us more."

A staff member focused on the practice of hospitality within the belief system of the Sisters of Mercy, but she emphasized that what is important about hospitality is not its link to religion, but that it offers "a welcome." While the interviewee was careful to note that she believes this ritual to be independent of religion, the feeling it generates is comparable to the feeling expressed by the participants and staff mentioned above. Though the focus on hospitality and the ritualized welcome may appear on the surface as a non-religious phenomenon, some have simply defined religious activities as those that generate what Émile Durkheim calls "collective effervescence" (1995). For Durkheim, religion is seen as a primarily social phenomenon wherein the collective energy from a gathering of individuals takes on a character of its own: a social euphoria. Religion builds community, but more than that, with the generation of great emotional or affective energy, individuals de-individuate and come to collectivize themselves. Recalling Carter's description of the ritualized writing of positive notes that were placed into the tires, the energy at WInC during the highly social event of the building's construction exemplifies this phenomenon. The rituals that propel affect through individuals and the social body—the welcome, the consecration of the building with positive messages, the labyrinth walks and morning meditations held for years at the Center—build ecological consciousness through intersubjective associations.

Victor Turner theorized ritual as a socially significant performance that creates meaning from the ambiguity of cultural liminality (1995). Understanding ritual as a mechanism of social cohesion may be particularly important for the study of a community that largely comprises immigrants striving to build a new home on the northern side of the border; indeed, his description of liminality shares much in common with what Anzaldúa describes as the borderlands. At one point, the Center's website bore this parallel out, citing Anzaldúa's claim that "The U.S.-Mexican border *es una herida abierta* where the Third World grates against the first and bleeds" (1999, 25). Anzaldúa saw the border as an in-between space populated by outsiders and from which residents can develop a critical consciousness—*la conciencia de la mestiza*—that draws on experiences of outsiderness to forge a new kind of belonging. In this case, the ritualized welcome and construction efforts act as place-making practices that invite participants to belong to the Center, to the community, to the land in Anthony. Reclamation of this land, as earlier activists sought to reclaim the Chicano/a homeland of Aztlán and Anzaldúa sought to create a new borderlands home, is an under-theorized strategy for ecological justice.

Although I have been writing about a connection between spirituality and ecological consciousness that uses affect to connect people to each other and to place as a means of belonging (i.e., spirituality can facilitate dwelling), spirituality is also en-natured at the Center. A longtime participant explained that before she joined, "I was used to staying at home with my family, praying in front of an image or something. Now we do it in nature, giving thanks to God for the day, for what we see, and to appreciate what we see because sometimes we are walking through the desert or the street and you don't appreciate what you are seeing, what nature is showing to you." She describes religious practice at the Center as en-natured, as an appreciation that joins us with nature and spirit others in a web of connectivity. For her, nature lives as much in the street as it does in the desert. Importantly, she does not claim that women are essentially linked to the natural world or that they share a unique connection to the spiritual world. This appreciation needs to be taught and what is most important, like others above have said, is the *feeling* that connects minds, bodies, spaces, and spirits. This is not an easy task as she went on to say: "We can talk about spirituality, but I cannot move your feelings inside." She is referencing the fact that some activities such as the sing-along to "Yo soy mujer" and morning meditations—spiritual activities that have affective import—have disappeared with the departure of the Sisters of Mercy; in their place the Center moved toward an increasing focus on personal and economic self-sufficiency. There need not be a discrepancy between practices that use affect to build social cohesion and those that instrumentally emphasize skill building (as many of the prior examples have shown). However, this example makes a case for diverse programming that includes a celebration of en-natured spirituality as part of individual and community healing.

Conclusion: lessons on affect and bioregionality

The shift from an overt spiritual presence to a latent one coincides with a larger shift from affective modes of gathering to combat social isolation toward gatherings explicitly aimed at education, skill building, and economic self-sufficiency. Examples of these programs include: the Small Business Academy and workshops on how to market one's paintings, food, jewelry, and clothes; workshops on how to fill out a job application and practice for an interview; and classes on how to study for and successfully complete the citizenship exam. This programming is important to women and makes a real difference in their lives and those of their families. While there is a critique of this model for the ways it fits within a neoliberal economic agenda that measures women's empowerment by their ability to participate in traditional economic and state-making practices, here I will note that the model of self-sufficiency is essential to disenfranchised women in an underserved rural town in New Mexico. Furthermore, a focus on economic self-sufficiency programming is more legible to national and regional funding agencies that can provide financial support for the Center's programming as well as its expansion into satellite offices in neighboring rural communities, which remains crucial for many residents who lack transportation. This point cannot be overemphasized since the recession led to staff reductions and the need for creative partnerships with regional service providers. Yet, in shifting away from programming that provides less visibly tangible rewards, WInC may lose that element of affective cohesion that was so important through its founding years. Staff that have been with the Center since the early days say that participants continue to find the sociality the best part of their experience, though women who have seen the evolution of WInC remark that the feeling is not as strong as it once was. This chapter argues that the ability to draw affective connections between human, nature, and spirit others is what pushes participants from instrumental education and community building toward the long-standing and deeper connections that have enabled the Center to become as strong as it has and to form a base for wider community change.

WInC's performative strategies to rework subjectivity in relation to one's body, to the natural world, and to the spirit world can create social, political, and environmental healing; they can be the basis on which to build an eco-feminist, decolonial community that is inclusive of both human and nature others. The ecological consciousness that emerges at WInC conceptualizes human/nature/spirit co-being, which gives more agency to non-humans than much of the literature on environmental justice allows since the focus is often on the effects toxicity and dislocation have on particular human communities rather than on the human–nature relation itself. It adds to ecofeminist models for green community because it recognizes the specific needs of working-class Mexican-American women, including the importance of developing a sense of belonging to the land, region, and nation. This chapter also contributes important lessons about ecological consciousness to environmental studies. It demonstrates the connection between raised consciousness and action from feminist and critical pedagogy studies; for example, learning about nutrition led to action—planting an organic garden. It also shows how place facts and systems awareness illustrated at WInC weave together social and environmental needs; for example, recycled tires provide an environmental good—a well-insulated Center—and a social good—happier, more comfortable participants.

Note

1 Modified from *Ecological Borderlands: Body, Nature, and Spirit in Chicana Feminism*. Copyright 2016 by the Board of Trustees of the University of Illinois. Used with the permission of the University of Illinois Press.

References

Anzaldúa, Gloria. 1999. *Borderlands/La Frontera: The New Mestiza*, 2nd edition. San Francisco: Aunt Lute Books.

Bingley, Amanda. 2003. "In Here and Out There: Sensations between Self and Landscape," *Social & Cultural Geography* 4(3): 329–345.

Carter, Mary. 2009. "Reflections: Women's Intercultural Center First Annual Report to the Community." Anthony, NM: Women's Intercultural Center.

Davies, Bronwyn. 2000. *(In)Scribing Body/Landscape Relations*. Walnut Creek, CA: AltaMiraPress.

Dodge, Jim. 1990. "Living by Life: Some Bioregional Theory and Practice." In *Home! A Bioregional Reader*, ed. Andruss et al. Philadelphia: New Society.

Durkheim, Émile. 1995. *The Elementary Forms of the Religious Life*. New York: Free Press.

Freire, Paulo. 1970. *Pedagogy of the Oppressed*. New York: Continuum.

King, Ynestra. 1983. "The Eco-feminist Imperative." In *Reclaim the Earth: Women, Speak Out for Life on Earth*, ed. L. Caldecott and S. Leland. London: Woman's Press.

"Mission." N.d. *Women's Intercultural Center.* www.womensinterculturalcenter.org/mission (accessed March 1, 2011).

Morris, Marla. 2002. "Ecological Consciousness and Curriculum," *Journal of Curriculum Studies* 34(5): 571–587.

Plumwood, Val. 1986. "Ecofeminism: An Overview and Discussion of Positions and Arguments," *Australian Journal of Philosophy* 64: 120–138.

Rowe, Aimee Carrillo. 2005. "Be Longing: Toward a Feminist Politics of Relation," *NWSA Journal* 17(2): 15–46.

Thayer, Robert. 2003. *Life Place: Bioregional Thought and Practice.* Berkeley: University of California Press.

Turner, Victor. 1995. *The Ritual Process: Structure and Anti-Structure.* Chicago: Aldine Transaction.

Tweit, Susan. 1995. *Barren, Wild, and Worthless: Living in the Chihuahuan Desert.* Albuquerque: University of New Mexico Press.

USDA Economic Research Service (Alisha Coleman-Jenson et al.). 2014. *Household Food Security in the United States.* www.ers.usda.gov/media/1565415/err173.pdf (accessed April 1, 2015).

10 The politics of land, water and toxins: reading the life-narratives of three women oikos-carers from Kerala

R. Sreejith Varma and Swarnalatha Rangarajan

India's narrative of nationhood and development in the post-independence years has largely remained truncated due to a variety of factors like the weakening of the tribal society, uneven patterns of growth, patriarchy and caste domination which have contributed to the lowering of women's status and their access to natural resources. Vandana Shiva (1988, 55) equates this decline of women to the death of the feminine principle 'Prakriti'—the creative feminine principle responsible for the co-creation of the world in tandem with 'Purusha,' the masculine principle. However, the complex linkages between gender and the environment are to be located in diverse historical and socio-political conditions and cannot be reduced to a simple ethic of care and loving rooted in women's reproductive capacities or women's way of knowing and being in a state of harmony with the world. The advocacy of ecofeminism in the Indian context of environmental justice activism can be largely seen as the attention on an oikos-household continuum where women's participation has been instrumental in regenerating these spaces under severe environmental and economic stress (Krishna 2009, 333).

The writers C. K. Janu (1970–), Leelakumariamma (1948–), and Mayilamma (1940–2007), whose life-narratives are taken up for analysis in this chapter are from Kerala, the southernmost state of India. The Kerala 'model of development' was hailed as a success since the statistical quality of life indicators put Kerala in the league of high-income developed countries in contrast to its counterparts in the low-income world. According to Bill McKibben (2007), Kerala offers hope for the future of the Global South given its low rates of infant mortality, high rates of literacy, and falling birthrates. He (2007, 123) approvingly notes that, "Kerala could be as significant a schoolhouse for the rich world as for the poor." However, Kerala's escalating environmental problems are numerous: the death of rivers due to deforestation, construction of dams, water extraction, discharge of effluents from industries, indiscriminate aerial spraying of pesticides and fertilizer use leading to changes in the diversity of species and organisms, poisons seeping into food chains and causing serious physical deformities in humans, uncontrolled sand mining hindering natural purification of water and the lowering of water levels in wells and lakes, a burgeoning eco-tourism industry that compromises the fragile coastal ecosystem and the values and traditions of indigenous communities. Besides raising questions about the sustainability of the Kerala model of development, this grim dossier of facts mirrors the environmental problems of India and the Global South and drives home the truth that development must be predicated not just on statistical indicators but also on equity, environmental justice and wisdom in using natural resources.

The life-narratives of these women writers coming from ravaged, subaltern environments demonstrate the écriture of resuscitation that brings the oikic home spaces into the public spaces of representation. They question interventionist technologists that harm the health of the land, rivers, and forests and clearly outline how women and nature are harmed when traditional ways of life are disrupted. Their narratives strive to make visible the invisibility of attritional "slow violence" (to use Rob Nixon's term)—namely the narratives of violence done to the earthscapes and the disenfranchised subalterns which erupt painfully and insidiously over a protracted period of time far below the level of visibility since they do not have the element of sensationalism which guarantees center-stage news coverage. Their ecofeminist ethics springs from the desire to restore their oikos to the exercise of 'active seeing,' which is a special kind of witnessing or attention that performs the advocacy function of raising public awareness and ensuring political action. These life-writings which at the local, bio-regional level map the disease of the physiological and planetary also become synecdochical texts mirroring the larger nodal network of what Heather Houser (2014) refers to as "ecosickness" narratives. In her eponymous work, Houser (2014, 6) points out that pervasive effects of technology ventures, ecological and social colonization have posed important questions for writers in the realm of the narrative affect: "how do interventions into the very stuff of life make us feel?" Houser (2014, 7) believes that this narrative effect portrayed through emotions like discord, fear, and anxiety acknowledges the multiple enmeshments of the body in the macroprocesses of environmental manipulation and leads to the awareness that emotions are the driving forces behind environmental ethics and political action.

The life stories of these women writers from Kerala employ the language of emotion to articulate the harm done to the fabric of the tribal and agro-centric ways of life that cannot be captured by information stacked in data charts. These earthcarers (to use Carolyn Merchant's term) are involved in local struggles and are on the periphery of the mainstream elite discourses on environmental activism. Their life-narratives offer concrete examples of how human subjects and environments are entangled thereby becoming what Iovino and Opperman (2014, 6) refer to as the "middle place" where matter enmeshes in the "discursive forces of politics, society, technology." These life-narratives offer unique examples of what Scott Slovic (2008, 28) refers to as "narrative scholarship" in which the human and other-than-human come together in a complex net of intra-being. They also offer unique Global South instances of what Carolyn Merchant (1995, 222) calls "partnership ethics" by "making visible the connections between economic systems, people, and the environment in an effort to find new economic forms that fulfill basic needs, provide security, and enhance the quality of life without degrading the local or global environment."

Kerala has been the emergence place of a number of women crusaders in service of the environment. A pioneering voice is that of Sugathakumari (1934–), the poet-environmentalist who spearheaded the iconic Save Silent Valley movement in the 1970s to save some of the oldest and biodiverse forests from the ravages of a planned hydroelectric project. However, environmentalism in Kerala has progressed beyond the wilderness protection drive of the 1970s, labeled as a middle-class, intellectual elite's concern for nature (Devika 2010, 765). The roots of the environmental movement in Kerala are firmly anchored in environmental justice movements which have seen the emergence of significant women earthcarers who engage in their unique brand of environmental justice activism to save the oikos—their home and hearth. C. K. Janu, Leelakumariamma and Mayilamma, whose life-narratives are taken up for analysis in this chapter are representatives of the grassroots 'oikos-carers' and women activists in Kerala and they redefine the space of environmental agency by positing a nature–culture continuum in which the awareness of the biological and social systems in which we are embedded is seen as a vital tool of survival and sustainability.

C. K. Janu's *Mother Forest* and the tribals' land struggles in Kerala

C. K. Janu's *Mother Forest* belongs to the wave of multiethnic writing emerging from the fringes of the Global South where nature does not represent pristine wilderness untouched by human presence. For the "empty-belly environmentalism" (Guha and Martinez-Alier 1997, xxi) that *Mother Forest* represents, the forest is as much a source of resources vital for human sustenance as it is a place for human/tribal–nonhuman animal cohabitation. Thus, the overarching question that the life-narrative of C. K. Janu, a tribal activist hailing from the Wayanad district of Kerala, raises is that of 'dwelling,' or rather the lack of it, and the poor material conditions faced by the Kerala tribals in the face of deforestation and expropriation of the tribal land by the non-tribals. The Kerala tribals who traditionally lived off the land were thrown into abject poverty when the migrated non-tribals encroached upon their lands and appropriated, through acts of persuasion and coercion, the majority of the land owned by the tribals. G. S. Jayasree (2012, xvi) usefully indicates that the "innocent" tribals "did not even have the notion land was 'property' and that they could be alienated from it. To the Adivasis, the first inhabitants of the land, there was no question of acquiring individual rights over the land. The land was not separable from their sense of collective identity; they were one with it and celebrated this union in all rites of passage." According to statistics in 2002, 24.7 per cent (i.e., 22,491 families) of Kerala tribals is landless and 34.02 per cent (or 30, 981 families) have less than an acre of land, making, as Bijoy (2006, 4332) indicates, "the overwhelming majority landless or near-landless."

The twofold division of *Mother Forest* represents the logic of binaries—the personal space as opposed to the public. As the translator N. Ravi Shanker (2004, xii) observes, "[The first chapter is] closer to Janu's inner world, while the second chapter [is] more polemical and belong[s] to the outer world." The first chapter contains descriptions of Janu's life in the forestscapes of Wayanad district, intimate vignettes of feudalism and the agricultural work done by the tribals and Janu's careful steps toward gaining literacy. The second chapter traces the changes that have come over the tribal community, Janu's participation in the Communist Party's activities, her later disillusionment with it as well as her early attempts at land agitations in places like Mananthavaadi and Munnar that mark her rise as an activist. Thus, the bipartite division of the text parses Janu's life into the dichotomized realms of *oikeion* and *politikon*. Jean-François Lyotard (2000, 135) explains that *oikeion* encompasses women, the children, the servants and notes:

> [E]verything that can be called "domesticity" in the old Latin sense, that which is in the *domus*, like the dogs, for example. In the final analysis, *oikeion* is everything that is not [public]. And the opposition between the *oikeion* and the *politikon* exactly matches up to that between the secluded on one side and the public on the other. The political is the public sphere, while the *oikeion* is the space we call "private", an awful word that I'm trying to avoid in saying "secluded".

However, the *oikeion*, in the tribal context, has an added complexity since, for the forest-dwelling tribals, the notions of 'domesticated' and 'wild' hardly exist. C. K. Janu (Bhaskaran 2004, 3) describes her life in the midst of "unending forests" and everyone gathering, as it gets dark, in the courtyard and spending "hours listening to what the forests mumbled." For the tribals who "lived in and with nature" (Kapoor 2004, 44), the lessons in ecological literacy were obtained early and at first hand:

> Nature lay open for little ones to learn from. Time and seasons could be told from the chirping of certain birds. The months could be counted when the leaves fell from the trees. From the darkening clouds descending on the hilltops and the forests we could gauge the direction of the wind. Our lives were so strongly interlinked with Nature, the earth and the trees. (Bhaskaran 2004, 51).

The deep topophilic connection that Janu as a girl experiences toward the land and 'mother' forest as documented in the first chapter is juxtaposed with the growing estrangement and topophobia in the younger generation of the tribals. In the opening of the second chapter, Janu (Bhaskaran 2004, 30) observes: "Mother Forest had turned into the Departmental Forest. It had barbed wire fences and guards. Our children had begun to be frightened of a forest that could no longer accommodate them."

C. K. Janu's *Mother Forest* is an example of "New forest texts" (Rangarajan 2016) or "Nava-Aranyakas,"[1] a term coined by Anu T. Asokan (2014) to analyze Mahasweta Devi's tribal texts as revisionist forest texts. Asokan (2014, 46) notes: "*Aranyakas* or the Vedic meditative texts (dated from 700–600 BC) . . . offer a mystical view of the forest. Traditionally, *Aranyakas*, an important limb of the *Upanishads*, were an important aid to philosophical reflection for men who retreated into the wilderness in preparation for the ultimate act of renunciation – '*Sanyasa*.'" However, 'Nava-Aranyakas' are not "wisdom texts . . . but volatile spaces in which texts of turbulence are written" (Asokan 2014, 47). Shorn of the metaphysical tinge, these 'New forest texts,' bring to the foreground a bricolage of themes which uncover the subsistence and survival-based issues of Indian environmentalism faced by forest dwellers who have been traditionally dependent on local natural resources (Rangarajan 2016).

The "livelihood ecology" (Gari qtd. in Martinez-Alier 2002, 10) that the New Forest texts like *Mother Forest* foreground reminds us that "antihuman environmentalism" (Nixon 2011, 5) will not work in the Global South context. Thus, C. K. Janu's deploring of deforestation and her "call for forest preservation" (Menon 2002) is as much about caring about nature as it is about attempts at "staying alive," to use the title of Vandana Shiva's 1988 book, as a community. Janu (Bhaskaran 2004, 55) perceptively asserts, "[M]any of our problems will be solved only if we continue to live and work close to the land."

With the changes in the traditional lifestyle of the tribals who largely depended on agroforestal and sustainable agricultural systems (Martinez-Alier 2002, 13), the majority of them turned themselves into mere wage laborers (Bhaskaran 2004, 30). The wages were not enough to keep hunger at bay and they did not have access to the forest which once provided them with a "bellyful of food" (Bhaskaran 2004, 7). The tribals helplessly witnessed the disintegration of agricultural lands that were allowed to lie fallow by its paper-owners, complaining that "agriculture was not paying" and often used to build concrete houses. The loss of land and access to the forest brought the tribals in greater contact with the civil society. Janu highlights the instances of sexual exploitation of young tribal girls by the 'outsiders' who took the girls to tribal hostels in towns in the name of education. Noting the transformations coming over the tribal community, Janu (Bhaskaran 2004, 48) comments:

> [The girls] imbibed only the wrong aspects of civil society. The way they spoke and the way they behaved became a matter of shame and degeneration . . . Unable to study or to pass the tests in new syllabi they lost their balance. They had to, [*sic*] for the cities to get their menial jobs done. Our people began to apply for such menial jobs. They became good-for-nothings by writing competitive tests and failing miserably in them. And the government ridiculed us further with figures that proved that our people were in a condition to compete with people from civil society. Certain ruling forces and power centres emerged who could stamp this society underfoot as a group of people who always failed.

Thus, uprooted from their lands and the forest environment and unable to quite blend in with the civil society, the tribals became tragic liminal figures or "nowhere people" (Nair 2003).

Janu (Bhaskaran 2004, 55) observes that all land struggles by the tribals in Kerala "have been struggles to establish the ownership rights of the real owners of this land for the right to live on it." Though *Mother Forest* does not cover the watershed Muthanga agitation on January 4, 2003 when thousands of tribals under the leadership of C. K. Janu and the Adivasi Gothra Maha Sabha (the grand assembly of tribals) encroached on the deforested portions of Muthanga and erected huts after the state government failed to keep its promise of land distribution to the tribals (Bijoy and Raman 2003, 1975–1976), it documents the circumstances that led to what climaxed in Muthanga. After describing a couple of land-occupations by the tribals led by her in places like Appootti, Vellamunda, Chinieru, and Kundara in Munnar, all of which were crushed using police force, Janu (Bhaskaran 2004, 54) wryly notes: "The only reason our people did not run away was that they had no place or land to run to. For that same reason they have been severely beaten up."

Echoing Ben Chavis's (qtd. in Sze 2002, 175) observation that "the issue of [socio-]environmental justice is an issue of life and death," Janu (Bhaskaran 2004, 55) asserts, "These were not just land encroachments. They were life and death struggles for our basic rights to live and die where we were born." *Mother Forest* is at once a place-based, topophilic narrative and a writing about displacement. It offers a gendered perspective on the state of 'uninhabitance' experienced by the Kerala tribals and rescues the Kerala indigenes from a long-haul "spatial amnesia" (Nixon 2011, 151–152). *Mother Forest* and C. K. Janu's socio-environmental activism should be understood, evidently, as part of the larger picture of growing struggles and writings for indigenous rights around the globe.

Leelakumariamma's Jeevadayini and Kerala's endosulfan disaster

The Green Revolution during the 1960s and 1970s was launched primarily to tackle the food problem in India through the introduction of advanced technology, high-yielding crop varieties, and an extensive use of pesticides. It was believed that by employing improved agrotechnology, the Malthusian prophecy of an impending population-resources imbalance could be overcome, and that material abundance could be achieved by combating scarcity through an unprecedented control of the environment (Shiva 1993, 14). India's agricultural experiment yielded remarkable returns in its early years; but the ecological ramifications of the Green Revolution soon became apparent as the chemical fertilizers impoverished the soil-fertility and insect pests developed immunity to the pesticides. Agriculture became more commercialized and the introduction of new seed varieties replaced the common traditional ones. Leelakumariamma's autobiography, *Jeevadayini* [*The Life-giver*], foregrounds the ambiguity of an agricultural officer who was raised in a traditional Gandhian farmer family in Kannur in northern Kerala, but had to campaign for the environmentally unfriendly programs of the Green Revolution as part of her profession. In her family, Leelakumariamma's mother, Pappiyamma, led the way by toiling hard in the fields using the traditional organic farming methods. Various crops like paddy, tapioca, *thina*, and *chaama*[2] were cultivated in their 15 acre farmland using cow dung as the main fertilizer for the crops.

While pursuing agricultural science in college, Leelakumariamma had to unlearn the traditional farming methods that she had picked up at home and learn the names of different chemical fertilizers and pesticides like Malathion and the need to produce high yields in agriculture. It is with deep contrition that she writes about her work at the Agricultural Office in Kasaragod district when she was entrusted to distribute to the local peasants chemicals like DDT and Sevin (which was manufactured by Union Carbide India Limited pesticide plant in Bhopal). Leelakumariamma also deplores the loss of many traditional varieties of paddy seeds after the introduction of Green Revolution's High Yielding Varieties (HYVs) of paddy. She had to face her colleague's antagonism when she revealed to the unsuspecting farmers that the pesticides they were taking home were poisons. At the end of Chapter Six, Leelakumariamma (Ajeesh 2011, 38) poignantly notes, "Within almost three years, we, Agricultural Officers, killed a rich heritage of the local farmers."[3]

Leelakumariamma's initiation into environmental justice activism was galvanized by the premature death of her elder brother Ramakrishnan, who, at her request, had come to stay with her to oversee the construction of her house in Periya village where she was transferred to. The livelihood of the villagers of Periya largely depended on the vast swathes of cashew orchards owned by the Plantation Corporation of Kerala (PCK), a public company run by the Kerala state government, where they either worked as laborers or obtained firewood. To ward off tea mosquitoes, PCK used to aerially spray the highly toxic endosulfan over the plantations using helicopters. Endosulfan belongs to the group of 'Persistent Organic Pollutants' (POPs) and is classified by "the US Environmental Protection Agency as a category Ib (highly hazardous) chemical, since it is easily absorbed by the stomach, lungs and through the skin" (Rajendran 2002). C. Jayakumar Thanal (2011, 88), the founding director of 'Thanal,' an environmental organization based in Thiruvananthapuram, lists cancer, various mental and physical disabilities, psoriasis, blindness, aborted pregnancies, as well as reproductive disorders as the common health risks associated with endosulfan exposure. The seriousness of the health impacts of endosulfan that Jayakumar Thanal illustrates throws into sharp relief the extreme callousness shown by PCK toward human and other-than-human lives in these villages by its rather criminal non-adherence to the standard norms prescribed in the instances of the aerial spraying of endosulfan. The stipulations with regard to the aerial spraying of endosulfan include:

1 The spraying of endosulfan should be publicly announced through microphones two days before.
2 Endosulfan should be sprayed only five-feet above the saplings.

3 Aerial spraying of endosulfan should be done only during early morning and late evening.
4 All water bodies should be covered before the spraying begins.
5 Cattle should not be allowed to enter the plantation up to twenty days after the spraying

(Ajeesh 2011, 43; Padre 2011, 47).

None of these steps was followed by PCK. As a result, the sprayed pesticide percolated down to the uncovered water bodies in the area, making them unfit for both domestic and agricultural consumption (Rajendran 2002, 2207). Crows and other birds killed after consuming the poisoned water soon became a common sight (Ajeesh 2011, 39) and Periya and other endosulfan-affected villages literally began to witness "the strange stillness" that Rachel Carson (1962, 2) described in the first chapter of *Silent Spring*. Leelakumariamma's brother fell ill within a year of his stay in Periya, his eyes frequently became watery and his sight also began to fail. Even doctors were unable to pin down the nature of his disease. The death of her perfectly healthy brother, whom she was very close to, just after a year of stay in Periya amply demonstrated to Leelakumariamma not only the physical health threats but also the emotional harm done by the pesticide use (Stein 2002, 194).

Leelakumariamma, who is credited as "the first person ever to come forward against the aerial spraying of endosulfan by the Plantation Corporation of Kerala" (Ajeesh 2011, 5), launched her crusade to end this 'toxic rain' by sending out petitions to the state's Chief Minister, Health Minister, Agriculture Minister, the District Collector and the Head of PCK in Kottayam. Her constant appeals to halt the spraying of endosulfan, however, did not incite any response. She, then, started to sensitize the local public about the dangers of endosulfan and sent ministers mass-signed petitions for three consecutive years (1994–1997). When she realized that it was pointless to wait any longer for governmental intervention in the issue, she decided to take the legal course.

Leelakumariamma's petition was considered by the Munsiff Court and it ruled for stay of the spraying of endosulfan in the plantations. The morning after the court judgment, PCK's Kasaragod district manager, along with a small group of people, came to Leelakumariamma's house and threatened her that she would suffer if she did not withdraw her complaint. Raising the gender question, Leelakumariamma (Ajeesh 2011, 50) asks if they would dare to turn up on her doorstep and threaten her if she were a male. Rob Nixon (2011, 145) has convincingly demonstrated how Rachel Carson and Wangari Maathai, two of the important female ecoactivists of the 20th century, "On all fronts, had to weather ad feminam assaults from male establishments whose orthodoxies were threatened by their autonomy." However, for Leelakumariamma, just as it was in the case of both Carson and Maathai, her "marginality was wounding but emboldening" (Nixon 2011, 145) and she, having the full support of the local villagers, decided, much to the astonishment of the PCK authorities, *not* to withdraw the complaint.

Although PCK moved to the Subordinate Court (which is superior to Munsiff Court), it also upheld the Munsiff Court's ruling. Following this, PCK approached the High Court of Kerala demanding Rs. 75 lakh as compensation from Leelakumariamma for the losses incurred by the company. At this crucial juncture of her campaign, Leelakumariamma was helped by T. P. Padmamanabhan, the Director of the environmental group called 'Society of Environmental Education in Kerala' (SEEK), who arranged Advocate Daisy Thampi to appear for Leelakumariamma in the High Court. After the hearing, the Kerala High Court asked PCK to wait for the Munsiff Court's judgment. On October 28, 2000, the Munsiff Court passed the landmark judgment ordering to stop the aerial spraying of endosulfan. Although she met with a life-threatening accident and is on crutches, Leelakumariamma continues her crusade for the ban of endosulfan across the country.

The damages done to the local ecosystem of these villages by twenty-four years (since 1976) of toxic exposure is immense. A study by Dr. V. S. Vijayan of Salim Ali Foundation (Mathew 2011), for instance, reveals "a decline in plant diversity between 40 and 70 percent, particularly for native species" in those villages where endosulfan was used. The toxic spraying of endosulfan has also left in its wake a very large disabled community. While endosulfan-related deaths in Kasaragod have crossed the 500 mark (John 2011, 14), the long-term casualties of the toxin are staggering. Y. S. Mohan Kumar (2011, 67), a Kasaragod-based physician who first associated the proliferation of health problems in the villages to endosulfan use, predicts that the after effects of the pesticide use are likely to haunt generations in Kasaragod for the coming fifty-sixty years! Rob Nixon (2011, 8) rightly points out that in the incidents of slow violence—like the endosulfan-impacted villages in Kasaragod—the "post is never fully post" since, as he (2011, 8) argues, "Violence, above all environmental violence, needs to be seen—and deeply considered—as a contest not only over space, or bodies, or labor, or resources, but also over time." Thus, with the "attritional lethality" (Nixon 2011, 8) of endosulfan lingering over these villages, there is no 'post-endosulfan' phase to speak of.

Leelakumariamma (Ajeesh 2011, 44–45) decided against fleeing the health-threateningly toxic atmosphere of Periya to other safer locations taking into account the plight of the indigent villagers who could not afford to do so. Leelakumariamma's ecofeminist ethics sees beyond the binary oppositions of privileged/oppressed and self/other which also renders the speciesist assumptions of human superiority erroneous. With regard to the menace of wild boars that come at night to feed on the vegetables grown in her backyard organic farm, Leelakumariamma (Ajeesh 2011, 25) maintains that the food crunch in the degraded habitats impels the wild animals to wander into village farms and asks, "What is the point in saving my crops by killing them?" Leelakumariamma's inclusive vision acknowledges, to cite Greta Gaard's (1993, 5) phrase, "sexism, racism, speciesism, and naturism (the oppression of nature) [as] mutually reinforcing systems of oppression" and gestures toward the "liberation of all oppressed groups."

The stories of toxic buildup, water pollution, deforestation, accelerated species loss and loss of habitats that form the narratives of 'slow violence' are rescued from invisibility and representational bias by women writing nature like Leelakumariamma in the Global South. However, it must be noted that Leelakumariamma's autobiography is as much "eco-protest" writing (Nayar 2014, 292) as it is a deep trauma narrative induced by environmental disturbance and personal loss. Her protracted battle to end the use of the toxic pesticide in her village underscores the typical Global South scenario where the matters of environmental impoverishment are embedded in "basic human rights and social justice" issues (Tarter 2002, 224; Comfort 2002, 229). The Kasaragod endosulfan disaster has been fodder also for creative writings in Malayalam like Ambikasuthan Mangad's acclaimed novel *Enmakaje* (2009) where Leelakumariamma makes a brief appearance and engages in a candid conversation with the novel's activist-protagonist Neelakantan about her anti-endosulfan campaign, and Francis T. Mavelikkara's play *Swapnangal Vilkkanunudu* (2013).

Mayilamma: Oru Jeevitham and the Plachimada struggle

Post-economic liberalization in India in the 1990s, the Coca-Cola Company, after an interval of sixteen years (1977–1993),[4] made a comeback to the Indian market and established, in March 2000, one of its bottling plants in the small village of Plachimada in the Perumatty Panchayat in the Palakkad district of Kerala. The company was set up in 34.64 acres in what previously were paddy fields and held out the promise of bringing 'development' to a 'backward' region (see Aiyer 2007, 643). The plant was intended to produce its popular brands like Coca-Cola, Fanta, Sprite, Limca, Kinley Soda, Maaza and Thumps-up and it started functioning in 2000–2001 (Bijoy 2006, 4333; Aiyer 2007, 643). The population of Plachimada, located on the borders of Tamil Nadu, comprises a predominantly dalit, tribal and a few Muslim communities dependent on the cultivation of crops like paddy, coconut, bananas, maize, etc. (Bijoy 2006, 4333). The prospect of jobs in the company was shattered as soon as the company started functioning as only a handful of the local people were given employment, citing their lack of education, in menial tasks like cleaning bottles. C. R. Bijoy (2006, 4333) describes the operations of the company thus:

> About 85 lorry loads of beverage products containing 550–600 cases each with each case containing 24 bottles of 300 ml capacity left the factory premises daily. Six bore-wells and two open-wells in the factory compound sucked out some 0.8 to 1.5 million litres of water daily. Within two years, the people around the plant experienced problems that they had never encountered before, the receding of the water table and the drastic change in the quality of water spread around 1 to 1.5 km radius of the plant. Water shortage upset the agricultural operations. Water became unfit for human consumption and domestic use.

Mayilamma: Oru Jeevitham,[5] the autobiography of Mayilamma, an unlettered woman belonging to Eravala tribe in Plachimada, begins with an episode where the uneducated tribal community comes to grips with the falling quality of drinking water and the different health problems posed by it. Children at the local Anganwadi Center, for instance, fall sick after consuming food prepared with the contaminated water, raising anxiety in their parents and the community at large. The water from the local sources which was traditionally used for drinking and cooking was found unfit for use and the village faced signs of impending drought. Contact with the contaminated water during dishwashing made hands itchy followed by a burning sensation. As ailments began to proliferate, the local tribals, who had previously avoided hospital treatment even at the time of pregnancy and childbirths, became frequent visitors of hospitals. With water scarcity and pollution beginning to upset the heath and the daily activities of the local population, the Panchayat started distributing water in tanker trucks. The water distributed by the Panchayat was no better either, as Mayilamma (Pariyadath 2012, 27) notes, the widespread rumors said that it was collected from the side of the Chittor River below a burial ground!

Vanadana Shiva (1988, 186), in her celebrated work *Staying Alive*, rightly emphasizes that groundwater is *not* a limitless resource but, "a critical part of the water cycle which depends on rainfall for its renewal." She (1988, 175) also affirms that "[v]iolence to the water cycle is probably the worst and most invisible form of violence because it simultaneously threatens the survival of all." According to the "Global Risks 2015 Report" by the World Economic Forum, "water crisis is the #1 global risk based on impact to society" (qtd. in "Facts about Water and Sanitation" n.p.). In the Global South context, women are the key players in the domestic water management (United Nations 2014) as well as the "water-providers" of the families (United Nations 2014; Shiva 1988, 171). Thus, disappearing water sources create new burdens and new drudgery for women (Shiva 1988, 171). Mayilamma (Pariyadath 2012, 23–26) makes note of the predicament of the tribal women in Plachimada who, after the water-mining of the Coca-Cola Company started, had to trek long distances in search of drinking water and wait carrying their pots in long queues even during the months of monsoon. She (Pariyadath 2012, 26) also indicates how it is often a trade-off between going to work and fetching water. For the underdeveloped rural tribal women, water-scarcity poses a concomitant risk of personal sanitation and hygiene owing to the non-existence of toilets and women's biological processes like menstruation and pregnancy. The issue of groundwater depletion in Plachimada was so severe that by 2005, the region was found to be a "water impoverished zone" (Aiyer 2007, 645).

The mass campaign against the Coca-Cola Company's water abuse in Plachimada was formally launched fittingly on Earth Day April 22, 2002 by the 'Coca-Cola Virudha Janakeeya Samara Samithy' (Anti-Coca-Cola People's Struggle Committee) with the participation of over 1,300 people, most of whom being the local tribals (Pariyadath 2012, 27; Bijoy 2006, 4334). The campaign was inaugurated by C. K. Janu who situated the Plachimada agitation within the larger picture of tribal protests for land and livelihood happening across the state which, Mayilamma admits, infused the gathered tribals with confidence. The tribals of Plachimada adopted the Gandhian 'satyagraha' style of peaceful sit-ins and 'hunger-strikes' to assert their water-rights and livelihood issues. The agitation, though facing indifference and opposition in its initial months, soon drew sympathies from environmentalists from across the country and the globe and Mayilamma became the very face of the agitation. The Plachimada struggle also garnered the attention of national and international press. *Mayilamma: Oru Jeevitham* provides the history of the Plachimada movement when the frail tribal-dalit women engaged in a Goliath–David kind of struggle against a global corporate power for water democracy and whose eventual victory[6] bolstered similar struggles against Coca-Cola happening elsewhere in the country and the world. The ecosickness narratives like *Mayilamma: Oru Jeevitham* also highlight how the environmentalism of the poor becomes "an environmentalism of survival and livelihood" (Ciafone 2012, 125–129).

The Coca-Cola Company took a lot of flak also for its rather hubristic act of handing out its waste sludge as fertilizers to the gullible local peasants. Mayilamma (Pariyadath 2012, 26) recalls that the waste smelled like decomposing carcasses. According to a BBC radio program exposé (qtd. in Bijoy 2006, 4334) dated July 25, 2003 on the Plachimada waste dumping:

> Of the three solid wastes analysed, one showed relatively high levels of two toxic metals, namely, cadmium and lead. Some other heavy metals, including nickel, chromium and zinc, were also present at levels significantly above those expected for background, uncontaminated soil and sludge. The presence of high levels of lead and cadmium is of particular concern. Lead is a developmental toxin in humans, particularly noted for its ability to damage the developing nervous system. Cadmium is especially toxic to the kidney, but also to the liver—it is classified as a known human carcinogen.

Clearly driven by the dangerous assumption that "natural resources and the poor [are] dispensable elements of ecosystems" (Shiva 1988, 213), the Coca-Cola Company's activities demonstrate the hierarchy of power relations and the burden of waste disposal that is differentially borne by those on the margins. Waste shifting from the Global North to the Global South, which is an objectionable practice from a transnational point of view, is also mirrored in this incident. In fact, Coca-Cola's entanglement in the issues of toxicity in the Indian context is more complex. The Center for Science and Environment (CSE) in New Delhi, for instance, published, in 2003, a report that indicated unacceptably high levels of pesticides like lindane, DDT, Malathion, and Chloropyrofs in the products of the Coca-Cola Company and PepsiCo (Summerly 2014, n.p.; Aiyer 2007, 644). The pesticide residues in these products were toxic enough to cause long-term cancer, damage to the nervous and reproductive systems, birth defects, and severe disruption of the immune system (Summerly 2014, n.p.). However, the Colas available in the U.S. were found to be pesticide-free (Ramesh 2004, n.p.). Furthermore, some newspapers reported that the farmers of Chattisgarh were successfully using Coca-Cola in lieu of costlier pesticides to protect their rice crops against pests (Putul 2014, n.p.). Ananthakrishnan Aiyer (2007, 646) observes that it is the reports of pesticide content in the soft drinks that marshaled the Indian urban middle-class opinion against the coke. On a related note, Mayilamma (Pariyadath 2012, 16), in her narrative, expresses her concern at the practice of some of her fellow tribals consuming toddy mixed with Coca-Cola for greater intoxication. Thus, Leelakumariamma's *Jeevadayini* and *Mayilamma: Oru Jeevitham* have the narrative of toxicity as the common horizon.

The tribal agitation against a multinational corporation (MNC) like the Coca-Cola Company should be situated in the broader history of tribal struggles in the state for land and livelihood. Mayilamma (Pariyadath 2012, 17–18) shares the history of Plachimada tribals' land-possession and land-distribution that involves painful stories of tribal exploitation. According to her story, the lands where the tribals toiled from dawn to dusk were taken away from them by the non-tribals by offering them a meager 100–200 rupees. The tribals were beaten up if they resisted. Mayilamma (Pariyadath 2012, 21–22) also documents a crisis situation faced by them not so long ago due to the unavailability of a proper burial ground. When someone died, the tribals had to cross the Kerala-Tamil Nadu border to come to the Tamil Gounders' land to inter the deceased. This issue sometimes escalated into violent conflicts between the two groups. The problem was finally solved when the district collector arranged a burial ground for the tribals at Nellimedu.

The tribals' loss of land is interlinked with the high levels of illiteracy in the tribal community. C. K. Janu (Bhaskaran 2004, 40), for instance, notes how the tribals lost their lands since they "were unable to find out or remember the survey numbers," thus, failing to prove their ownership. Landless and displaced, the tribal communities who were self-sufficient 'producers,' of grain were reduced to the state of 'consumers,' buying "everything from the shops" (Bhaskaran 2004, 52). Ananthakrishnan Aiyer (2007, 640) therefore usefully indicates that the Plachimada struggle also "needs to be analyzed as part of the unfolding agrarian crisis and not simply as a case of valiant community struggling against the rapacious practices of a transnational corporation like the Coca-Cola Company."

Conclusion

C. K. Janu, Leelakumariamma and Mayilamma fall under the category "grassroots women" whose socio-environmental activism is 'lived' rather than merely 'spoken' (Devika 2010, 759). They represent the breed of women ecoactivists who are on the margins of Kerala society like Seleena Prakkanam,[7] the female dalit leader of Chengara land struggle when over 7,000 landless dalit families occupied a part of Harrisons Malayalam Plantations in Chengara (Rammohan 2008, 15), and Pallikkal Bhavani,[8] a seventy-three-year-old resident of Padinjare Kallada "whose one-woman campaign against a sand-mining 'mafia' led to their packing up from the banks of the Kallada river" ("P.V. Thampi" 2010). As Mei Mei Evans (2002, 29) argues, personal testimonies like these are "the lifeblood of the environmental justice movement[s]" and the versions of environmental racism and ecological imperialism found in these testimonies draw attention to the government-corporate nexus that has engendered in laypeople "a web of mistrust" (Adamson and Stein 2002, 23) toward both the state and big businesses. Highlighting the subjection of 'human others' (women, people of color, [and in the Indian context, dalits], children and the poor) and "earth Others" (animals, forests, the land) (Warren 2000, 1) in a globalizing world, the life-writings of these Global South women activists throw light on the lives of communities willfully forgotten from the mainstream national discourse. These narratives call for a 'human-sensitive environmentalism' since the environmental and the social are closely imbricated in the Global South context. These life-writings highlight the "vaporized dwelling" of Indian tribals who, though being the original inhabitants of the land, suffer from a cultural contempt and marginality and being looked at as "inconveniencing anachronisms in a globalizing economy" (Nixon 2011, 153–164). These writings also offer resistance to forms of narrative violence employed by mainstream discourses which engage in the act of silencing crucial information and suppress important local stories so that they do not turn into judicial enquiries, monetary compensation or pieces of legislation.

These ecotexts as a whole fundamentally question the land, water, and agricultural commercialization and underscore the importance of concerted grassroots resistances against the same. These texts also seem to advocate what Vandana Shiva (2006, 1) calls, "Earth Democracy," which conceives the idea of "earth family," that is, the essence of the Indian saying "vasudhaivakudumbakam." Armed with slogans like "Our world is not for sale," "Our water is not for sale," "Our seeds and biodiversity are not for sale," 'Earth Democracy' responds to the strategies of privatization by corporate globalization (Shiva 2006, 2). Elaborating on the "insane ideology of privatization," Shiva (2006, 2–3) further adds, "Privatization encloses [for instance] the water commons. The enclosure of each common displaces and disenfranchises people which creates scarcity for many, while generating 'growth' for the few. Displacement becomes disposability, and in its most severe form, the induced scarcity becomes a denial of the very right to live." Though Shiva (2006, 3) specifically mentions the "victorious" Plachimada movement as located "at the heart of the emerging Earth Democracy," the ten key principles of Earth democracy justify the inclusion of the anti-endosulfan movement led by Leelakumariamma and the tribal land struggles like the Muthanga movement led by C. K. Janu—both conducted in a non-violent manner—in Earth Democracy. This is particularly evident in the fourth principle, which states that:

> All members of the earth community, including all humans have the right to sustenance—to food, water, to a safe and clean habitat, to security of ecological space. Resources vital to sustenance must stay in the commons. The right to sustenance is a natural right because it is the right to life. These rights are not given by the state or corporations, nor can they be extinguished by state or corporate action. No state or corporation has the right to erode or undermine these natural rights or enclose the commons that sustain life.

Inclusive of "diverse species, faiths, genders and ethnicities," Earth Democracy (Shiva 2006, 5) holds out the promise of "hope in the time of hopelessness." The life-narratives of the three women ecoactivists analyzed in this chapter also share this hope and optimism, for, as Linda Hogan (qtd. in Stein 2002, 200) perceptively put it, "Those who protested were the ones who could still believe they might survive as a people."

Notes

1 'Nava' in Sanskrit means 'new.'
2 Thina and Chaama are two varieties of millet.
3 All translations used in this chapter are ours, unless otherwise mentioned.
4 As Ananthakrishnan Aiyer (2007, 653) explains, "The Coca-Cola Company left India in 1977 after refusing to accept the terms of the Foreign Exchange Regulation Act, which reduced foreign ownership and equity to 40 percent in companies that produced consumer goods."
5 *Mayilamma: Oru Jeevitham* is transcribed by Jyothibai Pariyadath and uses an oral form to capture the flavor of tribal speech in Mayilamma's narration.
6 Mayilamma's life-narrative, although not documenting the episodes like the shutting down of the bottling plant in 2004, ends with an optimistic Mayilamma conveying her hope that the movement will meet with success.
7 Seleena Prakkanam's autobiography is titled *Chengara Samaravum Ente Jeevithavum* [*The Chengara Struggle and My Life*] (2013).
8 Pallikkal Bhavani's life-narrative is titled *Chorinte Manamulla Cheru* [*Sludge that Smells like Rice*] (2010).

References

Adamson, Joni and Rachel Stein. 2002. "Environmental Justice: A Roundtable Discussion with Simon Ortiz, Teresa Leal, Devon Pena, and Terrell Dixon." In *The Environmental Justice Reader: Politics, Poetics and Pedagogy*, ed. J. Adamson, M. M. Evans, and R. Stein. Tucson: University of Arizona Press.

Aiyer, Ananthakrishnan. 2007. "The Allure of the Transnational: Notes on Some Aspects of the Political Economy of Water in India," *Cultural Anthropology* 22(4): 640–658.

Ajeesh, T. 2011. *Jeevadayini*. Kottayam: DC Books.

Asokan, Anu T. 2014. "Environmentalism of the Poor: An Ecocritical Reading of the Selected Fictional Works of Mahasweta Devi." PhD diss., Indian Institute of Technology Madras.

Bhaskaran. 2004. *Mother Forest: The Unfinished Story of C.K. Janu*, trans. N. Ravi Shanker. New Delhi: Kali for Women & Women Unlimited.

Bijoy, C. R. 2006. "Kerala's Plachimada Struggle: A Narrative on Water and Governance Rights," *Economic and Political Weekly* 41(41): 4332–4339.
Bijoy, C. R. and K. Ravi Raman. 2003. "Muthanga: The Real Story: Adivasi Movement to Recover Land," *Economic and Political Weekly* 38(20): 1975–1982.
Carson, Rachel. 1962. *Silent Spring*. New York: Mariner Books.
Ciafone, Amanda. 2012. "If 'ThandaMatlab Coca-Cola' Then 'Cold Drink Means Toilet Cleaner': Environmentalism of the Dispossessed in Liberalizing India," *International Labor and Working-Class History* 81: 114–135.
Comfort, Susan. 2002. "Struggle in Ogoniland: ken Saro-Wiwa and the Cultural Politics of Environmental Justice." In *The Environmental Justice Reader: Politics, Poetics and Pedagogy*, ed. J. Adamson, M. M. Evans, and R. Stein. Tucson: University of Arizona Press.
Devika, J. 2010. "Caregiver vs. Citizen? Reflections on Ecofeminism from Kerala State, India," *Man in India* 89(4): 751–769.
Evans, Mei Mei. 2002. "Testimonies." In *The Environmental Justice Reader: Politics, Poetics and Pedagogy*, ed. J. Adamson, M. M. Evans, and R. Stein. Tucson: University of Arizona Press.
"Facts about Water and Sanitation." n.d. *Water.org*. http://water.org/water-crisis/water-sanitation-facts/
Gaard, Greta. 1993. "Living Interconnections with Animals and Nature." In *Ecofeminism: Women, Animals, Nature,* ed. G. Gaard. Philadelphia: Temple University Press.
Guha, Ramachandra and J. Martinez-Alier. (1997). *Varieties of Environmentalism: Essays North and South*. London: Earth Scan.
Houser, Heather. 2014. *Ecosickness in Contemporary U.S. Fiction*. New York: Columbia University Press.
Iovino, Serenella and Serpil Oppermann. 2014. *Material Ecocriticism*. Bloomington: Indiana University Press.
Jayasree, G. S. 2012. "Introduction." In *Kocharethi: The Arayar Woman*, by Narayan, trans.Catherine Thankamma. New York: Oxford University Press.
John, Babu. 2011. "Endosulfan durithabadithareSahayikkuka." In *Endosulfan Bheekaratha,* ed. B. John. Thiruvananthapuram: Chintha.
Kapoor, Dip. 2004. "Indigenous Struggles for Forests, Land, and Cultural Identity in India: Environmental Popular Education and the Democratization of Power." *Counterpoints* 230: 41–55.
Krishna, Sumi. 2009. *Genderscapes: Revisioning Natural Resource Management*. New Delhi: Zubaan.
Kumar, Mohan Y. S. 2011. "Oru Doctorude Sankadangal." In *Endosulfan Bheekaratha*, ed. B. John. Thiruvananthapuram: Chintha.
Lyotard, Jean-François. 2000. "Ecology as Discourse of the Secluded." In *The Green Studies Reader: From Romanticism to Ecocriticism*, ed. L. Coupe. London: Routledge.
Mangad, Ambikasuthan. 2012. *Enmakaje*. Kottayam: DC Books.
Martinez-Alier, Joan. 2002. *The Environmentalism of the Poor: A Study of Ecological Conflicts and Valuation*. Cheltenham, UK and Northampton, MA, USA: Edward Elgar.
Mathew, Roy. 2011. "Endosulfan Destroyed Biodiversity of Kasaragod Villages." *The Hindu*, April 22. www.thehindu.com/news/national/kerala/endosulfan-destroyed-biodiversity-of-kasaragod-villages/article1717767.ece (accessed September 21, 2015).
Mavelikkara, Francis T. 2013. *Swapnangal Vilkkanundu*. Kottayam: SPCS.
McKibben, Bill. 2007. *Hope, Human and Wild: True Stories of Living Lightly on the Earth*. Minneapolis: Milkweed Editions.

Menon, Usha. 2002. "When the Lost Soil Beckoned: Life Sketch Narrated by C. K. Janu," *Samyuktha* 2(2): 127–143. http://samyukta.info/site/book/export/html/290 (accessed 23 September, 2015).

Merchant, Carolyn. 1995. *Earthcare: Women and the Environment*. New York: Routledge.

Nair, Sindhu. 2003. "Janu and Nowhere People." Review of *Janu: C.K. Januvinte Jeevithakatha*, by Bhaskaran. *Infochange India.org*, August. http://infochangeindia.org/human-rights/books-a-reports/janu-and-the-nowhere-people.html (accessed September 23, 2015).

Nayar, Pramod K. 2014. "Indigenous Cultures and the Ecology of Protest: Moral Economy and 'Knowing Subalternity' in Dalit and Tribal Writing from India," *Journal of Postcolonial Writing* 50(3): 291–303.

Nixon, Rob. 2011. *Slow Violence and the Environmentalism of the Poor*. Cambridge, MA: Harvard University Press.

Padre, Sri. 2011. "Endosulfan Samaram." In *Endosulfan Bheegaratha,* ed. B. John. Thiruvananthapuram: Chintha.

Pariyadath, Jyothibai. 2012. *Mayilamma: Oru Jeevitham*. Kozhikode: Mathrubhumi Books.

Putul, Alok Prakash. 2014. "Farmers Tackle Pests with Colas." *BBC News*. November 3. http://news.bbc.co.uk/2/hi/south_asia/3977351.stm (accessed September 21, 2015).

"P.V. Thampi Award for Eco-activist." 2010. *The Hindu*. November 3. www.thehindu.com/todays-paper/tp-national/tp-kerala/p-v-thampi-award-for-ecoactivist/article866236.ece (accessed September 21, 2015).

Rajendran, S. 2002. "Pesticide Spraying in Kerala: Human Cost and Environmental Loss," *Economic and Political Weekly* 37(23): 2206–2207.

Ramesh, Randeep. 2004. "Soft-drink Giants Accused over Pesticides." *The Guardian*. February 5. www.theguardian.com/world/2004/feb/05/india.randeepramesh (accessed March 6, 2017).

Rammohan, K. T. 2008. "Caste and Landlessness in Kerala: Signals from Chengara," *Economic and Political Weekly*, 43(37): 14–17.

Rangarajan, Swarnalatha. 2016. "Women Writing Nature in the Global South: New Forest Texts from Fractured Indian Forests." In *Handbook of Ecocriticism and Cultural Ecology*, ed. H. Zapf. Berlin: De Gruyter.

Ravi Shanker, N. 2004. Translator's Note to *Mother Forest: The Unfinished Story of C.K. Janu*, by Bhaskaran. New Delhi: Kali for Women & Women Unlimited.

Shiva, Vandana. 1988. *Staying Alive: Women, Ecology and Survival in India*. New Delhi: Kali for Women.

——. 1993. *The Violence of Green Revolution: Third World Agriculture, Ecology and Politics*. London and New Jersey: Zed Books.

——. 2006. *Earth Democracy: Justice, Sustainability and Peace*. London: Zed Books.

Slovic, Scott. 2008. *Going Away to Think: Engagement, Retreat, and Ecocritical Responsibility*. Reno: University of Nevada Press.

Stein, Rachel. 2002. "Activism as Affirmation: Gender and Environmental Justice in Linda Hogan's *Solar Storms* and Barbara Neely's *Blanche Cleans Up*." In *The Environmental Justice Reader: Politics, Poetics and Pedagogy*, ed. J. Adamson, M. M. Evans, and R. Stein. Tucson: University of Arizona Press.

Summerly, John. 2014. "Pepsi and Coca-Cola Used as Pesticide in India Because They're Cheap and Get the Job Done." *Preventdisease.com,* May 6. http://preventdisease.com/news/14/050614-Pepsi-and-Coca-Cola-Used-As-Pesticide-In-India.shtml (accessed March 6, 2017).

Sze, Julie. 2002. "From Environmental Justice Literature to the Literature of Environmental Justice." In *The Environmental Justice Reader: Politics, Poetics and Pedagogy*, ed. J. Adamson, M. M. Evans, and R. Stein. Tucson: University of Arizona Press.

Tarter, Jim. 2002. "Some Live More Downstream Than Others: Cancer, Gender and Environmental Justice." In *The Environmental Justice Reader: Politics, Poetics and Pedagogy*, ed. J. Adamson, M. M. Evans, and R. Stein. Tucson: University of Arizona Press.

Thanal, Jayakumar C. 2011. "Endosulfan Charithram." In *Endosulfan Bheegaratha,* ed. B. John. Thiruvananthapuram: Chintha.

United Nations. 2014. "Gender and Water." www.un.org/waterforlifedecade/gender.shtml (accessed March 6, 2017; last modified October 23, 2014).

Warren, Karen J. 2000. *Ecofeminist Philosophy: A Western Perspective and What It Is and Why It Matters*. Lanham: Rowman& Littlefield.

11 Ecofeminism and the telegenics of celebrity in documentary film: the case of Aradhana Seth's *Dam/Age* (2003) and the Narmada Bachao Andolan

Reena Dube

On July 28, 2015 the Narmada Bachao Andolan (Save the Narmada Movement, hereafter NBA) turned thirty years old. The history of the struggle stretches back to the 1979 Narmada Water Tribunal decision to build 300 major, 350 medium, and 3,000 small dams on the Narmada river as well as to raise the height of the Sardar Sarovar dam from 80 meters to 120 meters, which culminated in the 2000 judgment of the Supreme Court of India to allow the dam construction to go on. Most recently, on June 12, 2015 the Narmada Control Authority gave the Gujarat government permission to raise the height of the Sardar Sarovar dam by another 17 meters, from 121.92 meters to 138.72 meters, which would mean that another 2.5 lakh people will face the same fate as the 320,000 people already displaced by the dam, according to unofficial estimates.[1] For a social movement that failed to achieve its main objective to stop the construction of the Sardar Sarovar dam across the Narmada river, the movement continues to garner great interest and respect across the world.

According to most commentators, the global significance and success of the NBA campaign issues from its innovative strategies of resistance that operated simultaneously at the grassroots, national, and international level. As such, the campaign's significance as a social movement extends far beyond India's national borders.Balakrishnan Rajagopal, a leading international scholar on development and social movements and a long-time observer and researcher of the Narmada struggle, notes that globally the NBA is "regarded as one of the signature public contestations of the twentieth century that redefined the terms of development, democracy and accountability" (Suyoggothi, 2013). Paul Routledge, assessing the NBA for his essay on transnational political movements, asserts that the NBA stands "as an emblematic example of the resistance to marginalization brought about by neoliberal development" (2008, 347).

In this chapter I examine the intersection between ecofeminist resistance and the discipline of film studies, specifically in the use of the medium of documentary by the Narmada Bachao Andolan. The movement has the unique distinction of having spawned a remarkable number of documentaries.[2] The social anti-dam protest movement occasioned by the Sardar Sarovar dam project qualifies as an ecofeminist struggle not only because the two most well-known faces associated with the movement are women—Medha Patkar, the social activist and founder of the Narmada Bachao Andolan in 1989, and the novelist Arundhati Roy, the 1998 Man Booker prize winner for the novel *The God of Small Things*[3]—but also because women have played central roles in the campaign. The NBA's struggle against the Maheshwar Dam in the state of Madhya Pradesh, for instance, has been led by the Narmada Shakti Dal, a separate women's organization within the NBA that was set up on March 8, 1988, International Women's day, and comprises female villagers from Maheshwar.

A second, equally important reason the NBA qualifies as an ecofeminist struggle is because of its use of non-violent Gandhian methods: organizing peaceful marches and protests, *dharnas* (sit-ins), hunger fasts, and an innovative "non-cooperation movement"in the Narmada valley against the payment of taxes that sought to deny entry to the villages to all government officials, except teachers and doctors.Their marches, demonstrations and even their water protest had frequently drawn violent reactions like lathi-charges and arrests from the government. In emulation of other disadvantaged groups which had successfully moved the Supreme Court,the NBA has been inspired to take up litigation against the current government for contravening the binding norms of the Narmada Tribunal and the judgments of the Supreme Court.

Thirty years after the launching of the NBA, and almost as long since the first documentary was made on the subject, Sashi and Mathur's *A Valley Refuses to Die* (1988), it is time to examine why and how the NBA struggle offered such a rich text of the theoretical issues endemic to ecofeminist struggles in postcolonialism when media is used to engage in globalization from below. I suggest that the use of the alternative media of documentary by women filmmakers in the NBA is both innovative and instructive. I examine one such example, Aradhana Seth's BBC-funded *Dam/Age* (2003), and how it negotiates with the power relations between center and periphery, especially in its engagement with global media in terms of telegenics of the celebrity personality cult. The purpose is to study the innovative evolving role of the documentary in the context of ecofeminist struggles in postcolonial developing societies wherein the traditional role of the documentary—the medium that captures, par excellence, the real, the ordinary, the present—undergoes transformation as a tool for the citizen/investigative journalist/popularizer in global transnational media solidarity campaigns.

Indian documentary and the nation state: a brief overview

In order to assess the current status of Indian documentary, we must begin by recognizing that the medium of documentary, more than any other audio-visual medium, has a symbiotic relationship with the nation state. In the Indian case, documentary served the longest apprenticeship under its inherited colonial ideology. The majority of documentary films made in India during roughly the first three decades after independence fit squarely within what is known as the Griersonian tradition: artisanal practices and imagery borrowed from the European avant-garde of the 1920s, the "voice of God" narration, the predictable problem/solution format, the objective being to discourage direct political action on the part of the audience by convincing them that a benevolent socialist-leaning state has the means and initiative to solve all problems. The independent documentary movement in India, pioneered by the likes of Anand Patwardhan, rejected the Griersonan tradition and consistently de-emphasized the aesthetics of practice in favor of activism.[4] In his documentary about the NBA, the *Narmada Diary* (Simantani Dhuru/Anand Patwardhan 1995), started as a project on behalf of the NBA, Patwardhan and Dhuru comment ironically on the Grierson tradition by juxtaposing contemporary news footage about the struggle to save the Narmada river valley from being destroyed by a proposed hydroelectric dam with the soundtrack from an FD documentary "Village of Smiles," in which the narrator quotes Prime Minister Nehru's famous statement that "Dams are the temples of modern India" (Ghosh 2009, 81).

It is commonplace that moments of crises or revolutionary change often lead to exciting and spectacular leaps in audio-visual mediums such as documentary film, lending it a sense of urgency and immediacy, which often translates in terms of documentary realism and political relevance. The Emergency (1975–1977) under Prime Minister Indira Gandhi functioned as one such crisis which radically questioned the citizenry's belief in the project of the independent nation for architects of the modern documentary like Patwardhan who took the nation state as the main frame of reference of his critical gaze, constituting a discourse which defined the political credentials of a film according to whether it was affirmative or critical in its address, and the ideal of the filmmaker as a revolutionary citizen of the nation who still acted on behalf of it. This discourse was, however, a counterpoint to the political and social imaginaries proposed by the social and political movements that were gaining traction post-Emergency, and which critiqued the lack of alternative visions of the nation state (Wolf 2013, 366).

Throughout the 1980s it was the women's movement and its issues which led people to start making documentaries and which constituted much of its audience. Yet the discursive atmosphere was experienced by some women filmmakers as being as restrictive as the state propaganda form which they were seeking to oppose, "we have got stuck with the form of socialist realism without the environment of revolution—that bred that form in the first place" (Bhushan 1998). Therefore women's films which explored the real and the political through the fictional, the theatrical or popular instead of social realism were critiqued for not being political enough. In addition the standardization of the process of NGO-ization led to documentary film abandoning open-minded inquiry and reflection through aesthetic exploration and turning into vehicles for visualizing an argument or didactic position This state of affairs prevailed right through the 1990s up to the early 2000s, and was, according to Wolf, "a period of discontent, which despite or because of this paved the way for the lively documentary scene that we find today" (2013, 367–368).

In contrast, I offer a reading of a documentary made by a women documentarian in the early 2000s that challenges the scenario described by Wolf. Admittedly the documentary is occasioned by a revolution in the form of a global socio-political movement against big dams and the neoliberal development. In order to fully appreciate how a woman filmmaker like Aradhana Seth opens up the documentary format to a rather unusual inquiry and reflection through aesthetic exploration, I begin with two questions suggested by Wolf in her survey and summary of the current documentary scene in India that are particularly relevant to *Dam/Age*. The first asks how constitutions of the *self* intertwine with the shifting textures of the real, the people, the nation state and its discontents. And the related question concerning the framing of the subjects of documentary as protagonists through processes of othering or how they create and imagine themselves in front of the camera? The second inquiry is about how the development of documentary languages can be addressed while acknowledging local embeddedness as well as transnational affiliations (Wolf 2013, 362).[5]

We obtain very interesting answers, I suggest, if we examine the former question in terms of the latter in the context of the Seth documentary. That is to say that given the transnational nature of the documentary, we find that in Seth/Roy's documentary *Dam/Age* (2003), the classic documentary categories and distinctions of the self, the real, citizenship, the process of othering and local embeddness are muddied up.

Aradhana Seth's *Dam/Age* (2003): the making of a global celebrity woman activist

Dan Brockington begins his analysis of the rise of celebrity within development studies with a caveat, "I must first dispel a common misconception. The study of celebrity is not a shallow or trivial exercise . . . we need to explain its popularity and success" (2014, 88). Brockington sees an increase in the importance of celebrity advocates of development issues as poverty expands in the Global North where celebrity industries have been present at the same time as we see the expansion of celebrity industries into the Global South with internet connections, media industries, and mobile phones. Implicit in Brockington's (2008) forecast about the expansion of the role of celebrity advocates in development issues is the assumption that the traffic will be one-way, celebrities from the North visiting and advocating development issues affecting the Global South. But Seth/Roy's artfully constructed documentary *Dam/Age* (2003), aligning the story of the fight against the Sardar Sarovar dam alongside the personal narrative of novelist and essayist Arundhati Roy's engagement with the NDA, records the making of a transnational woman celebrity activist from the Global South, and therefore breaks new ground in Indian documentary filmmaking. I use the construction 'Seth/Roy' as the makers of the film not only because Roy performs herself as a celebrity activist and is the voiceover and narrator of the entire film, but by Seth's own admission it is a film made "with" Arundhati Roy, "in every sense of the word"; therefore it is Seth/Roy's clever and careful interlacing of political issues and personal drama that makes of the film a record of the making of Arundhati Roy into a translocal celebrity activist.

Although it is true that by the time of the documentary Roy had already published her critical essays on the developmental politics in "Power Politics" (2002c), of nuclear testing in "The end of Imagination" (1998) and big dams and in "The Greater Common Good" (1999), and was already known on the international lecture circuit, her ascent as the global celebrity activist was occasioned by her personal engagement with the NBA movement in terms of both her donation of $30,000 Booker Prize money and the Rs, 20 lakhs royalty for her essay, "For the Greater Common Good," and in no less measure by her participation in the anti-dam NBA demonstrations and rallies.

Roy's activism has come in for a fair share of criticism particularly by scholars and academics who critique Roy's alliance with the Narmada Bachao Andolan (NBA) for not being deeply immersed in that struggle alone, that her targets are diffuse, that her facts merge into fiction, that her non-fiction is often ill-informed and self-serving. For instance Julie Mullaney, critically examines Roy's "particular brand of ecofeminism" which positions her as "*the* public voice of India's anti-globalization movement" (2002a, 63, 57), and concludes that it undermines both the history of such protest, as well as the role of noteworthy activists and scientists like Vandana Shiva. Like Mullaney, Routledge notes the irony of Roy's critique of the global marketplace even as she "herself [is] the beneficiary of a globalized literary market" (2008, 345). Graham Huggan is more explicit in his criticism of Roy's manipulation of her celebrity image through what he describes as her "*strategic* exoticism," her "incorrigibly photogenic" appearance, and her "life story containing almost as many carefully leaked secrets as [*Small Things*]" (2001, 77). Roy has also received negative criticism from social and political environmentalists, like Ramachandra Guha, who claim that Roy's writing about the Narmada Dam Project is "self-indulgent and hyperbolic," her "vanity . . . unreal," her anti-dam essay ("The Greater Common Good") replete with "signs of self-absorption" (2000).

In contradistinction I suggest that the key to Roy's success as a global celebrity activist is precisely the qualities so maligned by the literary and academic establishment. In order to argue this, I examine Roy's performance in *Dam/Age* of her transnational activist identity and show how at each turn in the film the Seth/Roy combine makes strategic choices in order to represent Roy as a popularizer, a translator, a facilitator, and a celebrity rather than a scholarly, academic expert. The process begins right from the opening shot of the film with Roy's iconic profile at dusk on a boat gazing out over the Narmada landscape. Roy is foregrounded, she stands above the sitting woman and man protestors/passengers on either side of her. This shot functions as a historical placeholder—indicating her early (she sports her earlier hairstyle) engagement with the movement, and is a sequence that is repeated again and again, especially in a flashback 40 minutes into the narrative. Roy's voiceover reads from her text "The Greater Common Good" about crossing the river, climbing up the bank and laughing at the ludicrousness of the Supreme Court's concern that displaced children have parks in which to play, as the camera shows us the shoreline, skyline, and the waters of the rushing river from Roy's perspective on the boat finally ending on an interior shot of Roy in the present moment as she reads from her essay on camera. The self-referential opening of the NBA documentary, especially the laughter, prompted Ramachandra Guha (2000) to characterize it as "a straight lift from the first lines of that monument to egotism, Ayn Rand's *The Fountainhead*." I suggest it is precisely these kinds of easily discernable allusions to popular literary icons from the Global North, as well as the self-referentiality and the laughter (which occurs again and again in the film), that Roy uses to great effect to construct her translocal celebrity activist identity.

The process of foregrounding Roy continues as the subtitles appear on screen, "In October 2001, the Supreme Court of India charged Arundhati Roy with contempt of court. This is the story of the events that lead to her arrest. *Dam/Age: A Film With Arundhati Roy*." Thus we are skillfully prepared by Seth/Roy to hear Roy on Roy occasioned by a film ostensibly about the damage caused by the age-of-dam: specifically to be acquainted through Roy on the damage caused by big dams through the damage suffered by Roy because of her engagement with the anti-dam movement in the age-of-dams.[6]

Given Seth/Roy's easy assumption of the self-referential mode from the opening moments of the film, it comes as a surprise when 28.5 minutes into the film, in the segment where Roy is preparing for the court date, a monologue occurs about this being a public-political fight yet "it's also personal." Roy confesses to finding it "very difficult . . . to decide when to be personal and when to be public. In a moment like this, maybe you don't want to put your personal self out there, but you have to, because it can only be fought in the public arena. And it must be done." The mystery is dispelled if we read this discussion as Seth/Roy embodying the popular feminist principle of the personal is political. Similarly little more than 13 minutes later once again Roy foregrounds her femaleness during the disclosure that in addition to the Supreme Court judgment day on March 6, there is a hearing in Kerala on February 15 to ban *The God of Small Things* for the last chapter presumably corrupting public morality. Roy quips, "It is as though it's that time of the month," and laughs, then explains somberly and despondently, "For it to be coming up right now is strange, because it is as if the legal net is closing round me in some ways." The quip and laughter foregrounds Roy's identification with "that time of the month" for all woman when they feel overwhelmed, and with her despondency she shares the vulnerability of a courageous woman.

In the second segment of the film Roy is framed inside the Supreme Court in interaction with her lawyer, Prashant Bhushan, in his chambers; her voiceover elaborates the charges against her. That is when we see the dam affected people for the first time in the film, in the distance in long shots as they shout slogans while walking through fields and on the receding river banks, as the camera on the boat pulls away and Roy's voiceover commences, explaining the nature of her association with the NBA. The connection between the Narmada and Roy, she says, is personal and child-like: "I owe rivers a debt of gratitude for all kinds of understanding. I grew up in Ayemenem on the banks of the Menachil. I always think that as a very young child, the hours that I spent catching fish were the hours that made me the writer that I am." Once again the initial iconic profile shot of Roy on the Narmada is repeated. Not only is the association with the Narmada located in her childhood memories of the town made famous by *The God of Small Things*, but also the description of her relationship echoes her writerly style in the novel where her language experimentation includes fluid adoption of children's thoughts and images: she says she understands the river not as an environmentalist or an ecologist, "but just because a river was my friend when I was little. And the loss of a river is a terrible aching thing."

Roy's singularity as celebrity writer develops through the next segment of the documentary when we witness a protest where Baba Mahariya, an indigenous tribal activist, speaks on camera in Hindi, and subsequently Medha Patkar, the other famous woman social activist associated with the NBA, appears for the first time. Patkar is indistinguishable in dress and appearance from the protestors and is filmed turned away from the camera, addressing the gathering in Hindi, as if in implicit acknowledgment that as the grassroots organizer and leader neither she nor the NBA are the focus of the documentary. Underlining the difference in identities, roles and linguistic communicative capabilities of the two women for the NBA, Roy is absent from this particular sequence except for a very brief glimpse of her amongst the cheering crowd of women audience members. Then Roy's voiceover commences over the faces of the gathered protestors, explaining that her decision to write about the Narmada was driven by her feeling that "what was missing in the fight was the story, the whole story, all the connections. And I really wanted to tell the story in all its details but accessible to the ordinary reader". Thus Roy introduces her evolving persona as an essayist and writer of social and political commentary. At this point it will be useful to suspend an examination of the narrative of the documentary in order to pursue the question of Roy's self-definition as it relates to her role for the NBA.

A global celebrity woman activist?

In the introduction of her essay, "The Greater Common Good" Roy reaffirms the idea that it was the story embedded in the NBA that drew her, even though at the time the fight for the NBA had "entered a newer, sadder phase"; she writes, "I went because writers are drawn to stories the way vultures are drawn to a kill. My motive was not compassion. It was greed. I was right. I found a story there" (2002b, 49). What is notable, however, is that Roy's declaration that she is a serious non-fiction writer, an expression of intent couched in phraseology one would expect from an investigative journalist, follows immediately after her deliberate recalling of her fiction writing: she ends her suggestion that the scale of the 21st century is the small and specific rather than big with these words, "Perhaps right now, this very minute, there's a small god up in heaven readying herself for us. Could it be? Could it *possibly* be? It sounds finger-licking good to me" (2001, 48–49).

Roy's linguistic strategy above is typical of her non-fiction writing: she constantly subverts expectations. Just when she seems to speak with authority about a subject, she takes refuge in her literary fiction persona; then when we adjust our expectations to hearing from a novelist about political matters, a writer-activist in other words, she lapses into the most pedestrian populist expressions (parodying an advertising or public relations campaign slogan), lending credibility to her claim to be a middle-class citizen-journalist. The net result is that her non-fiction writing is, I suggest, a postmodern pastiche that patches together a number of discourses, from the low to the high, without letting herself be held hostage by any particular one.

While much has been written about her play on language in her novel, little sustained analysis has been forthcoming on her non-fiction writing. The majority of the reactions veer to the extremes: from Guha labeling it self-indulgent to Steve Palopoli who, reviewing *Dam/Age*, gushes like a star-struck fan calling Roy "a fascinating muse for India's culture of protest, able to drop insights like 'The only thing worth globalizing is dissent' at any given moment" (2002a). Roy's stated views on language are at odds with her own practice in her non-fiction writing and therefore critical to understanding her refusal to accept the label of the writer-activist.

Precisely in the years when poststructuralism and postmodernism swept the academy and challenged the normative notion of language and the plain style with singularity of meaning and authorial intention and the literary world was celebrating the liberation into linguistic multiplicity and heterogeneity, Roy claims to eschew the playful postmodern style of her novel for a common sense view of language. She writes:

> As a writer, one spends a lifetime journeying into the heart of language, trying to minimize, if not eliminate, the distance between language and thought. 'Language is the skin on my thought,' I remember saying to someone who once asked what language meant to me.
>
> (2000, 135)

She espouses this view of language implicitly, even when she characterizes the discussions at the World Water Forum at The Hague, as "the ritualistic slaughter of language as I know and understand it . . . mask[ing] intent. . . .They breed and prosper in the space that lies between what they say and what they sell" (2000, xviii). Roy repeatedly ascribes this obfuscation in the language of experts:

in many of her essays of the time she questions the idea that only experts can speak on such urgent matters as nuclear war, privatization of India's power supply, and construction of monumental dams in India. As translator and popularizer of "the story" of the NBA the playful, creative Booker Prize winning novelist aspires to write in the "accessible" language for the "ordinary reader."[7] What is left unsaid, however, is that the ordinary readers she has in mind are middle-class global Indians like her and the transnational audience abroad. In an interview with Barsamian, Roy acknowledged that she says what other middle-class people like her would like to say but cannot. The middle class has a valid role in the leadership of the NBA because, as she puts it, it was middle-class engineers who designed the dam.

She elaborates this idea of being the envied celebrity spokesperson for the ordinary citizenry in the context of disputing the appropriateness of the appellation of the writer-activist. Roy describes the current global state of things for the writer: the writers are "free" to an unprecedented degree, more numerous and "commercially viable" than ever before, virtually "influential, wealthy superstars," "torch-bearers," their work the "benchmark" for art as it "is" and "ought to be." All of this has led, Roy writes, to what she calls the "almost farcical" scene in India where the "Indo-Anglian" writer has emerged as the new career pursued by ambitious parents, corporate advertising seeking endorsement and infotainment media. She notes that while "in the early days" she was introduced as the "almost freakishly 'successful'" author of *The God of Small Things*, she now "in twenty-first-century vernacular" is referred to as "a 'writer-activist'. (Like a sofa-bed.)" She explains her discomfort with the appellation even when used admiringly. She reasons that "the double-barreled appellation, this awful professional label," has been given to her, not merely because her work political:

> in my essays I take sides . . . What's worse I make it clear that I think it's right and moral to take that position and what's even worse, use everything in my power to flagrantly solicit support for that position. For a writer of the twenty-first century that's considered a pretty uncool, unsophisticated thing to do . . . uncomfortably close to the territory occupied by political party ideologues.
>
> ("The ladies have feelings, so . . ." 2002d, 175–176)

Roy has drawn yet another distinction between her non-fiction essays and other commentaries: her essays are not only written in the common language, but are also "flagrantly" political and thus infringe on the province of political "ideologues." The branding of her non-fiction writing as a "writer-activist" diminishes both writers and activists: the writer is seen as "too effete" for having the wherewithal to "publicly take a political position," and activists, the professional position takers, are seen as "crude, simple-minded, one-sided understanding of things." Her "fundamental" objection, however, is "that this attempt[s] to 'professionalize' protest" (ibid., 186).

She specifies that her own involvement is neither in order to raise causes nor because she is a writer or activist, but the as a "human being," a citizen reacting in writing to the immense political and social upheavals taking place because it "just happens to be the most effective thing a writer an do" (ibid.). She recommends the de-professionalization of public debates on matters affecting ordinary lives; to take back from the experts, "to ask, in ordinary language, the public question and to demand in ordinary language, the public answer" (ibid., 187). That is the reason that even as she admits that there has been a fair amount of writing on the subject, she characterizes most of it as "for a 'special interest' readership":

> Experts and consultants have hijacked various aspects of the issue . . . and carried them off to their lairs where they guard them fiercely against the unauthorized curiosity of interested laypersons . . . Disconnecting the politics from the economics from the emotion and human tragedy of uprootment is like breaking up a band. The individual musicians don't rock in quite the same way. You keep the noise but lose the music."
>
> ("The Greater Common Good" 2002b, 46–47)

It is precisely due to this role of the experts that Roy writes, "For the people of the valley . . . their most effective weapon—*specific* facts about *specific* issues in this *specific* valley—has been blunted by the debate on the big issues" (46). For Roy the role of experts is Brahmanical, to colonize knowledge and profit by it; she uses the saying "There's a lot of money in poverty" to imply that experts and scholars and indeed organizations like the World Bank are "parasite(s)" (188). What is happening in globalism lies, Roy clearly suggests, "outside the realm of common human understanding" and it is only artists practicing a "new kind of art" that can bring it within the realm of common understanding through humanizing the "boardroom speeches into real stories about real people with real lives." The virtue of her writing, she seems to suggest that it is both more political and less political in the traditional sense, because she has "no personal or ideological axe to grind. I have no professional stakes to protect". Her politics, if it must be identified, is humanist, that of a "citizen" ("The Ladies . . ." 2002d, 187). And while there can be no rules for a writer, the trajectory of a writer must not be weighed down with morality and responsibility; yet the refusal of morality and responsibility "can only lead to bad art." This is a "new" kind of politics: a politics not of governance but of resistance, of opposition, of forcing accountability. Roy ends the essay "The Greater Common Good" in her inimitable style, with a quotable quote, "In the present circumstances, I'd say that the only thing worth globalizing is dissent. It's India's best export" (2001, 191). And Arundhati Roy is our best global export of dissent, we might add.

The aesthetics of representing a media-savvy global celebrity woman activist

When we next see Roy in *Dam/Age* it is in the present framed in the tastefully appointed, softly glowing interior of her house; she explains, "I really believe that the story of the Narmada valley is the story of modern India. Not just of modern India but it is a story of what is happening in the world today." With this last statement, Roy has drawn even more starkly the difference between Patkar's and her own perceived role in the NBA. We might say that while Patkar's is a place-based 'militant particularism,' Roy's own engagement with the NBA may be characterized as 'global ambition.' The terms are those of David Harvey who, theorizing social movements, borrows a term from Raymond Williams to argue that place-based resistances frequently articulate a 'militant particularism.' But the ideals forged through this militant particularism's affirmative experience of solidarities in one place have the potential to become generalized and universalized as 'global ambition': a working model for a new form of society that will benefit all humanity. However, Harvey also notes that militant particularisms are profoundly conservative, grounded in the perpetuation of patterns of social relations and community solidarities. Therefore it stands to reason to wonder, as Harvey (1996) does, whether at the scale of global ambition, militant particularisms become impossible to ground, let alone sustain. It is precisely for this reason I suggest that Patkar and Roy seemed to have defined their roles differently with respect to the NBA. And as a result of which, the NBA achieved great success in forging a national and transnational coalition, working on innovative resistance strategies at multiscalar levels.

In less than four and a half minutes from the start of the film, Seth/ Roy have successfully accomplished three tasks in the making and marketing of a global celebrity woman activist: they have framed Roy's global ambition visually, linguistically, and discursively; foregrounded her difference from the people associated with the NBA, including Patkar, and finally overlaid the narrative of a people's movement with Roy's personal narrative. This overlapping of the two narratives occurs visually and discursively as Roy's explanation progresses: it occurs visually as Roy's voiceover discloses her realization that in writing the story of the Narmada valley, "instinct wasn't enough", that "I needed to know the algebra" and as she speaks the camera cuts from her in profile reading and marking a book, to a close-up panning shot of what appears to be words on a concrete wall, evoking the concrete structure of the Sardar Sarovar dam, except that this one is blue in color. Then as the camera pans slowly over the words, we realize we are looking not at a dam wall but at the book cover of Roy's *The Algebra of Infinite Justice* (2002a).[8] The discursive overlapping occurs, interestingly, with Roy's repeated use of the metaphors of "light" and "dark" to determine "what is happening in the world today": she uses the former phrase to describe the role of power, including her own, and the latter to designate the people who don't count, yet who pay the costs and suffer the collateral damage; later in the segment as she talks about how the centralization of resources through neoliberal development schemes such as big dam building benefits the rich and makes "[ordinary] people melt into the darkness and disappear." She uses the metaphor of light as she describes her own task as a writer, "to increase the area which is lit," and the metaphor is implicit in the space she professes to command, that is as celebrity to draw the media's light: "I suddenly realized I have the space, I command the space in which to raise a dissenting voice and if I don't do it, it is as political an act as doing it." The use of the twin metaphors appears to be a very deliberate strategy on the part of Roy and in doing so she appears unaware or uncaring of the colonialist legacy of the civilizing mission implicit in the metaphors. The effort is to distinguish herself from the experts and academics who might be exercised by the politics of it, but who do not write in an accessible manner or address for ordinary people, but to use language, as she memorably said in one of her early interviews on NPR in answer to an observation that her language use in the *The God of Small Things* was innovative, "Language does not use me, I use language!"

Roy's narrative about her present legal woes moves artfully back and forth between her activist identity and her persecution which individuates her and adds to her celebrity status: how false and "ludicrous" litigations against her, Patkar and Bhushan by five lawyers in the Supreme Court devolves into a suo moto contempt of court charge against her alone by the Supreme Court because of her affidavit questioning the wisdom of the Supreme Court in taking up a case that was dismissed by the police station. Similarly artfully Roy casts the significance of fighting the contempt of court charge on the double registers of a personal vindication and "political" litigation, against the contempt of court act: the lawsuit is "challenging" and opening to "criticism" an "unaccountable institution" (body? Roy's own?) as it "wades into public life" (the Supreme Court and Roy herself?) and is increasingly powerfully affecting the lives of millions of people. The segment ends with the first of the only two interactions between Roy and an NBA subaltern woman protestor: at 15:40, in the middle of a demonstration she tells the subaltern woman in halting Hindi, "What if they put me in jail?" "We will be there. We will fight." "I am going to be locked up on the 6th." "We'll see." "They say I am not apologizing." "Why? What's our crime? They build their dams, they flood our lands, they displace people, destroying everything. They should apologize, [Roy joins in] why should we?" "It's our right to ask questions," ends the woman, almost ventriloquizing Roy herself. Through the exchange Roy's individual legal trouble is embraced by the subaltern woman as "our crime" for which "we" will see, we will be there, and we will fight. Instead of the displaced people it is Roy who has become the cause célèbre, prefiguring the most icon-making scene of the film where in its final moments we see that the demonstrations and slogan shouting, instead of espousing the cause of the NBA, are now being undertaken to express solidarity with the woman celebrity activist.

The actual "story of the Narmada", begins almost 17 minutes into the film and ends 5 minutes later with the "Rally for the Valley." In a documentary film of 50 minutes, the "algebra" and the "mathematics" of the argument with the world about the Narmada, as Roy termed it, constitutes only one-tenth of total film time. On the one hand this decision on the part of Seth/Roy is surprising given the visual skill with which this "story" is told by Roy with the aid map, historical FD newsreel, official documentaries, actual footage of the lives of the displaced in the city slums and the submerged villages. The duo display the same skill in a later clip when the interconnections drawn by Roy between war, bombs, and dams echoes eerily in the speech about the achievements of the government (nuclear tests in 1998, Kargil war in 1999 and lifting of the stay on the Narmada dam in 2000) by the then Home Minister L. K. Advani. The overlapping fortuitously allows Roy to underline her own persecution, "when he makes the connections it's a patriotic act, and when I make the connections it's a criminal offense."

We begin to appreciate the wisdom of Seth/Roy's decision to substitute dry facts and figures when we are treated to the visual sumptuousness and emotional resonance of the famous 'Rally for the Valley.' Roy recalls her call for the 'Rally for the Valley'—"we called for people, for outsiders, for people from all over the world, and from people's organizations and resistance movements from all over India to travel through the valley to see what the struggle was all about." Note Roy's frequent use of the lists in her address to people. "Such enumeration," notes Bishnupriya Ghosh, "is typical of Roy's strategy, which turns multiple corporealized bodies palpable, immediate, and intimate for the transnational audience who religiously follow her untiring lecture circuit in the Global North, on television and radio, in print, and over the Internet as the chronicler of local struggles" (2011, 270–271).

Today we know that the "Rally" established Roy's sheer star power viz-à-viz the national and international media, activists, and volunteers. The various groups from across the world who visited the Narmada valley during the "Rally" to learn about the struggle, to participate, and to subsequently disseminate information about it, continued to maintain links with the NBA and conduct solidarity work on its behalf. For example, the group Narmada UK was formed after several individuals who had participated in the 1990 'Rally for the Valley' along the Narmada, and the subsequent PGA conference in Bangalore, decided to conduct solidarity work in the UK in support of the NBA (Routledge 2008, 346).

We watch the video of the rally alongside Roy as Medha Patkar walks past the camera along with the other protestors and Roy is shown standing up on stage, "The whole valley rose up—somehow the spontaneity, the happiness that someone *outside* cared, that their struggle had registered, that it mattered [emphasis mine]," she says as she watches the video of the Rally. Her words not only acknowledge her own outsider status, but also make explicit her ability to call upon other outsiders to care and to register. The segment ends with us watching Roy watch herself. Then comes Roy's second interaction in the film with a subaltern woman:

> That's Dedlibai of Domkhedi. At the end of the rally she gave me this basket of all the different kinds of seeds, and all the different kinds of plants

> and foods, and wheat and lentils and rice that they grow. And she is here saying, where will they be able to replace this for us if they throw us out of here, if they take the river away from us, if they take our ancestral land away from us, will they be able to give us all this?

Dedlibai is well known as one of the go-to cast of the tribal NBA woman activist characters that featured in a number of documentaries occasioned by the NBA, what Bishnupriya Ghosh calls "NBA docs."[9] Like her predecessor, the anonymous subaltern woman interlocutor in the first interaction with Roy, Dedlibai charmingly and willingly enacts her own subalternity while acknowledging Roy's 'outsider' celebrity status with her gift.[10]

Roy's poetic address to the subaltern old man Bhaijibhai, "the last person I met in the valley" who was not strictly a displaced person because he had lost 17 of his 19 acres to the "wonder canal," only adds to Roy's unique auratic identity:

> Bhaijibhai and his people forced to smile on government calendars, denied the grace of rage, squashed like bugs by this country that they're supposed to call their own. Bhaijibhai, Bhaijibhai, when will you get angry? When will you stop waiting? When will you say that's enough and reach for your weapons, whatever they may be? When will you show you show us the whole of your resonant, terrifying, invincible strength? When will you break the faith? Will you break the faith, or will you let it break you?

At first blush, Roy's poetic exhortation of Bhaijibhai appears to be an explicit call to arms. Not surprisingly then that later in the decade the Maoists expressed the desire that Roy serve as a mediator between them and the government. An offer that Roy refused, but agreed to write a tract about them, resulting in the essay "Walking with the Comrades" (2010).[11] Foreshadowing her response to the invitation by the Maoists seven years later, the call to arms to Bhaijibhai, when it is repeated, in part, 40 minutes into the film over a repeated iconic shot of Roy from the beginning of the film, it is crafted to enhance her transnational translatability:

> When I first went to the valley, I used to say that I am not here because *my* house is being submerged, or my fields are going under water. But my world view is being submerged, that's why I am here. But it's not just that anymore. They are knocking at my door, they are coming for me and therefore I know that I have to fight with all the skill I that I have, which is my words, my ideas, my ability to communicate. I believe that the only hope and the only thing worth globalizing is dissent. And I think that when the Supreme Court comes for us, for the artists, for the writers, for the filmmakers, for the musicians, we have to show them our terrifying strength we have to fight back with our art.

Notice the introduction of two new elements from the previous version addressed to Bhaijibhai: first is the use of phrases and meter that are reminiscent of the famous statement and poem "First they came. . ." written by Pastor Martin Niemöller (1892–1984) about the cowardice of German intellectuals and quoted in the U.S. Holocaust Museum. The second element is how disarmingly the militant element becomes translated into the arms that a writer has at her command: words and ideas.

Roy is ever mindful of her international audience and she expertly reels them in by direct reference as well as through topics that would resonate with them; for example, at a little under half an hour into the film, we find Roy addressing her American audience with her artless confession about her perceived image abroad, specifically America, "I'm always invited to portray myself as this radical person who is being hunted down by these institutions in this native banana republic . . . There are many things about India like a banana republic, but that goes for America as well." The confessional is followed by reference to the free market as an unwelcome third in her present "conversation" with an institution in the society that she inhabits, "this is the way in which democracy becomes more sophisticated." She then links up free market to the two dams on the Narmada—the Sardar Sarovar (state sponsored) and the Maheshwar (privately funded)—where the brutality and the corruption are the same, if not exacerbated in the case of the latter.

The introduction of Roy's reminiscences about "the dawn we marched through to capture the Maheshwar dam site" introduces just the right note of romance and activism back into the polemic:

> I can still remember the sound just the crunching of feet and the flip flap flip flap of slippers. We started at three in the morning. We walked for three hours: farmers, fisher folk, sand miners, writers, painters, filmmakers, lawyers, journalists—all of India was represented, urban, rural, touchable, untouchable. We were not just fighting against the dam, we were fighting for a philosophy, a worldview.

Roy loops back to the issue of the free market and controlling institutions, like the WTO and World Bank, that control even developing countries like India through their parliamentary institutions. The second to last sequence of this segment is an anecdote that although is ostensibly about her personal sense of her outrage at the World Bank President James Wolfensohn's visit to India, may as well have been crafted by the screenwriter Roy (an identification mentioned by Roy herself in this account) to introduce the racial viewpoint, that most essential ingredient of transnational politics:

> I was just seeing this white man in a pinstripe suit addressing the peasants of India all over again. I didn't want to hear what he said. I didn't want to hear his reassurances. I just kept thinking who the hell are you, how are you back here after all these years? Was all this for nothing?

> The World Bank pulled out of the SSP in 1993, but the mess they've made lives on.

Appropriately, the anecdote about the patronizing white male World Bank president is punctuated with a shot of Patkar and others in the *jalsamarpan* protest; this in turn leads to the final sequence of the segment where Roy draws the contrast as she recalls that in 1999 while the protesters stood in chest-deep water while their households and the homes were being washed away, at the same time the Supreme Court held three sessions to discuss whether Arundhati Roy had lowered the dignity of the press. She bursts into laughter as she reads the charges. The contrast and laughter brings the focus of the film back to Roy, this time via the Supreme Court's obsession with Roy herself. According to Laura Wright, this laughter marks the supplanting of the "victimization of valley residents" due to displacement to "Roy's victimization by the Supreme Court of India." The film is no longer preoccupied by the "fate" of the Narmada river and the affected people, Wright notes, "the narrative now focuses on the fate of Roy." In fact for Wright the last part of the film, what she refers to as "the third 'act' of Seth's film" (the first and second act being punctuated by Roy's laughter occasioned by the Supreme Court), is characterized by the silencing of the polluter who "pollutes the stream of justice." The story of Roy's persecution is mediated "through other, less coherent voices," writes Wright, and "the audience is aware of the chaos that exists without Roy's framing discourse" (2010, 125–126).

In contrast to Wright's reading, I suggest that far from being "silenced" or put "offstage," Roy's celebrity becomes mythologized and the transition of her transnational identity to a social and political mediator and translator gains even greater gravitas. This is expressed by Roy herself in two statements in the final moments of the film: she expresses the hope that her trials make of her, "a different kind of writer now," because having "written the book I wanted to write . . . now I am swimming in the river of life and I hope what comes out will be literature at the end of it." The second statement is Roy's concluding comment in the last moments of the film, "the establishment has always feared writers because they have the weapon of clarity" that, when used effectively, "can be deadly." The mythologizing itself takes place primarily visually through the first scenes in the film that show preparations for a protest, not for the NBA but for Roy herself who has by now completely displaced the NBA and in fact eclipsed it: from posters being painted, written, and stitched, in English and Hindi, to Roy perusing a fax with a long list of signatories the camera focuses on for a couple of seconds;[12] to Roy watching from the balcony of the Supreme Court with tears in her eyes, the protestors gathered and chanting below to walking through the chanting crowd with her legal entourage.

The final step in the mythologizing and iconizing of Arundhati Roy happens when Seth/Roy use camera in the way that is typical of guerilla cinema. That is to say Wright's "chaos," the "seemingly inarticulate crowd," and "the camera at a loss" in the minutes anticipating the judgment may in fact be read as a technique of "the imperfect" borrowed from revolutionary cinema. The imperfection in guerilla cinema is an acknowledgment that in terms of technical achievement in comparison to the mainstream, revolutionary cinema will always fail; and second, as committed art whose true goal is for the artful to disappear, revolutionary cinema cannot strive for disinterestedness. Thus for instance scenes of protest may be composed of short, handheld, and unstable shots, as they are filmed in the midst of the riot police and smoking tear gas canisters. Similarly the uneven and jerky camera movements and indistinct urgent voices we hear on the audio track as we await the Supreme Court's judgment in *Dam/Age*, impart a sense of immediacy and urgency to the images and experience. In importing this technique from revolutionary cinema for the ending, Seth/Roy's attempt is to elevate the film from a personal record into a political documentary.

Aradhana Seth and Arundhati Roy's film *Dam/Age* in its documentation of the transition of a glamorous multicultural or postcolonial fiction writer into a transnational celebrity woman activist, also exemplifies a new kind of experiment in women documentary filmmaking, one where new subjects are explored innovatively, without sacrificing aesthetics, and where the distinction between "activism and art" are blurred.[13] Through very careful crafting and choreographing of her image and her words, we see Seth/Roy position Arundhati Roy as a fascinating icon of India's culture of protest who refuses, as she herself puts it in "Power Politics," the luxuries of "subtlety, ambiguity, complexity" to ask "in ordinary language, the public question and to demand, in ordinary language, the public answer" (2002c, 24). Hence, it seems to me to be an oversimplification to call Roy the "voice" of the subaltern (Ghosh 2011, 270) or that "Roy does speak for the untouchable subaltern (in Spivak's sense) masses," (Wright 2010, 123).

A much more reasonable description, as I have attempted to show through an examination of Seth/Roy's documentary *Dam/Age*, is through her role as a transnational celebrity woman activist of globalized ambition, whose attempts to lay bare and universalize the working model of herself to benefit all of humanity, ultimately proves inimitable and ungeneralizable: she acknowledges as much when, at the end of the film, she chooses to pay the fine and refuses, in her own words, to martyr herself for a cause that is not hers alone. "Choosing to suffer," she says, "isn't exactly my style." And her style, we are compelled to conclude, is inimitable.

Notes

1 The most recent decision has been widely criticized as favoring Prime Minister Narendra Modi's home state Gujarat, at the cost of the lakh people in Madhya Pradesh and Maharashtra. See "The News Minute" (2014).

2 K. P. Sashi and Ratna Mathur's *A Valley Refuses to Die* (India, 1988), Ali Kazmi's *Narmada: A Valley Rises* (Canada, 1994), Simantini Dhuru and Anand Patwardhan's *The Narmada Diary* (India, 1995), Jharni Jhaveri and Anurag Singh's *Kaise Jeebo Re* (*How Shall We Survive*, India, 1997), Sanjay Kak's *Words on Water* (India, 2002), and Franny Armstrong's

Drowned Out (U.K., 2002) along with two shorter videos, Arvind Pillamarri's *I Will Report Honestly* (India, 1999) and Leena Pendharkar's autobiographical foray *My Narmada Diary* (U.S., 2002).

3 See Parsai (2003). Although the article talks about Roy's donation of the prize money from the U.S.-based Lannan Foundation Award to fifty remarkable people's movements, institutions, and individuals who are engaged in making India into a real democracy, it also mentions Roy's previous donation of the Booker award to the NBA.

4 John Grierson was a Scottish-born filmmaker and administrator who became a leading figure in the British documentary movement of the 1930s, when he headed the film unit at the Empire Marketing Board (EMB), an organization charged with creating propaganda that would promote trade among the colonies of the British Empire, including India. See Hanlon (2014).

5 Although I pick up on only her last two suggestions for this chapter, Wolf names five that deserve further study (2013, 362).

6 Aradhana Seth, the director, acknowledges this unique collaborative feature of the film, because although many had wanted to make a film with Roy, "she wasn't going for it, because she didn't know what it would turn out like." It was agreed that it would not be "only a conversation" with Roy, but rather a record of her thoughts, her writing, her words as the court event was unfolding, therefore "really a film *with* her, in every sense of the word that made the project come to fruition in the first place" (Palopoli 2002).

7 In her interview with Barsamian, Roy makes this clear: "I don't see a great difference between *The God of Small Things* and my works of nonfiction ... My whole effort now is to remove that distinction ... It is very important for me to tell politics like a story, to make it real" (Barsamian 2007).

8 In 2005 Roy dramatically turned down India's prestigious Sahitya Akademi Award for *The Algebra of Infinite Justice* (2002a) in a well-publicized media event. Roy chose instead to use the glare of publicity to make a symbolic gesture of protest against the Indian government's policies of targeting the disenfranchised.

9 Ghosh writes, "The 'NBA docs' (as I shall call them) shot on film and video have played a robust role in garnering global legibility for a series of heterogeneous struggles." See Ghosh (2009, 60).

10 Judith Whitehead provides some examples of this phenomenon of what I call enactment of subalternity and what she calls "naturally conservationist imagery" about the tribal during the 'Rally for the Valley' in the summer of 1999: "the arrival of noted writer and activist Arundhati Roy was accompanied by hundreds of journalist from both India and abroad including from *The Times of India, The Indian Express, the BBC,* and *The Hindu*. All had to be accommodated in the tiny hamlets of Jalsindhi and Domkhedi. For the first time since visiting the Narmada valley, I saw a group of male Adivasis attired in 'traditional' *lungis* and turbans, some playing long flutes and ready for the photographers. When I happened to step beside a group playing music, an irate photographer from *The Times of India* asked me to step aside, since I was spoiling his photograph of 'authentic' Adivasi culture." See Whitehead (2010, 154–155).

11 Koteswar Rao, better known by his *nom de guerre* Kishengi, the leader of the main guerilla force of India's Maoist extremists, fighting a violent insurgency against the Indian state telephoned the BBC from an undisclosed location to say that the Maoists would halt their campaign if the government invited intellectuals and rights activists like Roy to mediate in peace talks. Roy ruled out becoming directly involved in any talks, but told *The Guardian* that she considered the offer "serious" and said she felt "both sides should call a ceasefire."

She was "a writer, not a mediator. I don't think I would be very good at it," she said. "It's a serious responsibility and there are people who would be good at it," although she would consider being an "observer". See Burke (2010). Also see Roy (2010).

12 In the interview with Palopoli, Seth identifies the list as a "fax of support from Toni Morrison, Jonathan Demme, Woody Allen, Susan Sarandon, Robert Redford, Bernardo

Bertolucci," which said, "Dear Mr. President, we write to express our full support for the author Arundhati Roy and our concern in regards to the contempt petition against her. . . . We fully support the right of free speech" (Palopoli 2002).

13 In an interview conducted with Barsamian (2004), Roy is overtly critical of the documentaries that are produced in India, noting that "few of them transcend the boundaries between activism and art."

References

Barsamian, David. 2004. *The Checkbook and the Cruise Missile: Conversations with Arundhati Roy*. Cambridge, MA: South End Press.

——. 2007. "Interview with Arundhati Roy." *Progressive*, February. www.progressive.org/news/2007/07/5078/interview-arundhati-roy

Bhushan, Madhu. 1988. "Activism and Art: A Dialogue with Deepa Dhanraj." *Deep Focus* 1(3): 32–39.

Brockington, Dan. 2008. "Powerful Environmentalisms: Conservation, Celebrity and Capitalism," *Media, Culture and Society* 30(4): 551–568.

——. 2014 "The Production and Construction of Celebrity Advocacy in International Development," *Third World Quarterly* 35(1): 88–108.

Burke, Jason. 2010. "India's Maoist Extremists Ask Arundhati Roy to Mediate in Conflict with State." *The Guardian*, March 7. www.theguardian.com/world/2010/mar/07/arundhati-roy-maoist-mediator.

Ghosh, Bishnupriya. 2009. "We Shall Drown, but We Shall Not Move: The Ecologics of Testimony in NBA Documentaries." In *Documentary Testimonies: Archives of Suffering*, ed. B. Sarkar and J. Walker. New York: Routledge.

——. 2011. "The Politics of the Icon." In *Global Icons: Apertures to the Popular*. Durham: Duke University Press.

Guha, Ramachandra. 2000. "The Arun Shourie of the Left." *The Hindu*, 26 November 26. www.thehindu.com/2000/11/26/stories/13260411.htm

Hanlon, Dennis. 2014. "Making Waves: Anand Patwardhan, Latin America, and the Invention of Indian Third Cinema," *Wide Screen* 5(1): 1–24.

Harvey, D. 1996. *Justice, Nature and the Geography of Difference*. Oxford: Blackwell.

Huggan, Graham. 2001. *The Postcolonial Exotic: Marketing the Margins*. London: Routledge.

Mullaney, Julie. 2002. "Globalizing Dissent? Arundhati Roy, Local and Postcolonial Feminisms in the Transnational Economy." *World Literature Written in English* 40(1): 56–70.

Palopoli, Steve. 2002. "Village of the Dam: Director Aradhana Seth Shows a Side of India You'll Never See on TV in 'Dam/Age.'" *Metro Santa Cruz*, October 30–November 6. www.metroactive,com/papers/cruz/10.30.02/damage-0244.html

Parsai, Gargi. 2003. "Arundhati Roy Donates Rs. 1.67 –cr. Award Money." *The Hindu*, January 23.

Routledge, P. 2008. "Transnational Political Movements." In *The SAGE Handbook of Political Geography*, ed. K. R. Cox, M. Low, and J. Robinson. New York: Sage.

Roy, Arundhati. 2002a. *The Algebra of Infinite Justice*. London: Flamingo.

——. 2002b. "The Greater Common Good." In *The Algebra of Infinite Justice*. Delhi: Viking Books.

——. 2002c. "Power Politics." In *The Algebra of Infinite Justice*. Delhi: Viking Books.

——. 2002d. "The Ladies Have Feelings, So. . . ." In *The Algebra of Infinite Justice*. Delhi: Viking Books.

——. 2010. “Walking with the Comrades.” *Outlook Magazine*, March 29. www.outlookindia.com/printarticle.aspx?264738

Suyoggothi. 2013. “The Story of Narmada Bachao Andolan: Human Rights.” *Essential Thinkers*, October 3. https://essentialthinkers.wordpress.com/2013/10/13/the-story-of-narmada-bachao-andolan-human-rights/comment-page-1/

“The News Minute.” 2014. July 1. www.thenewsminute.com/technologies/56 Whitehead, Judith. 2010. *Development and Dispossession in the Narmada Valley New Delhi*. San Francisco: Pearson Education.

Wolf, Nicole. 2013. “Foundations, Movements and Dissonant Images Documentary Film and Its Ambivalent Relations to the Nation State.” In *Routledge Handbook of Indian Cinemas*, ed. K. M. Gokulsing and W. Dissanayake. London and New York: Routledge.

Wright, Laura. 2010. “Swimming in the River.” In *Wilderness Into Civilized Shapes: Reading the Postcolonial Environment*. Athens: University of Georgia Press.

Afterword

Izabel F. O. Brandão

The act of looking back offers us a possibility to reflect on what has been done, and ponder about our own perceptions as concerns a certain issue. We may, as Foucault (1984) claims, find ourselves in a heterotopia—a real place that can be seen in opposition with a utopian and unreal place. Border spaces exist between heterotopia and utopia which act as mediators in such places. The mirror—"a placeless place" (Foucault 1984, n.p.)—is the territory that sums up both spaces.[1] This is where the writing of this Afterword is located: as a mediator sensing a path that tells me that the connections between women and nature still have a long way to go before one can safely say that no conflict is to be found any more. This is to begin with; let feminism and ecofeminism[2] direct the connections that follow.

I had immense pleasure reading this book, for if, on the one hand, it tells me that people from different countries understand the organization of society along the lines that criticize this very organization because of its patriarcal values, which still oppress beings—humans, non-humans—and more specifically women; on the other, such criticism is not an empty one. All readings bring suggestions of how to construct a different world along a more egalitarian and hopeful line than what we see today. I understand the analyses that form the corpus of this book as a varied one, which commune in a range of perspectives that enters into dialogue, as poses Bakhtin (1984), referring to the multidirectionality of discourses (Bauer and Mckinstry 1991). Such is the idea of the complex intersectionality behind the articles which come from different fields of knowledge in their dialogue about gender, race/ethnicity, sexuality, class, the body, and their interconnection with nature.

My reading of the articles found at least three lines of thought which interweave, interconnect, and intersect the authors in their approach to ecofeminism, women, transgender individuals, animals, and nature: (1) ethics of care, and essentialism; (2) narratives of oppression as related to body images; and (3) social action, and activism as related to life-narratives, and other writings. This is in no way a suggestion that the articles have to follow a stark understanding, for their dialogue finds echoes elsewhere, as I intend to point out.

One of the relevant questions that permeates ecofeminist thought from the start is its interdisciplinary basis, rooted in its "philosophical plurality" and the theoretical debate allowed by such basis (Gaard and Murphy 1998, 4). Yet the many trends within ecofeminism open up the possibility of diversity as far as a problematic connection between women and nature is concerned, one of which is the source of many dissensions within feminism itself, that is, essentialism. In order to depart from a context inside ecofeminism, it is possible to come across many ideas, some of which support the connection whilst others contest it, turning the area into a territory of potential conflict, or as claim Gaard and Murphy (1998, 6): "Static categories that define our nature or individuals can only be distortions." On the one hand, if the relationship involving nature/culture/male/female/masculine/feminine/other may be quite worn, on the other, it still is far from an easy solution.

Ecofeminists such as Karla Armbruster (1998) situate the problem by calling into the fore other eco/feminist critics, like Ynestra King or Susan Griffin who work such an equation—woman–nature—within a perspective that vindicates a more liberatory connection. All humans, they claim, are driven by cultural forces that can blind them to their (our) participation in nature (Armbruster 1998, 100). Stacy Alaimo (2000), however, calls for a redefinition of nature, for this will also mean a redefinition of woman (we can say women). The argument used by Alaimo, supported by Val Plumwood, Rose Braidotti, and Donna Haraway, is that feminists (and ecofeminists) need neither to escape nor to associate with nature provided that a transformation of the gendered concepts such as nature, culture, among others, is effected.

The above debate leads to the fact that both women and nature are "socially constructed," and that nature, "a profoundly gendered realm," is "a site for many struggles for power and meaning" (13), as rightly claims Alaimo. In this sense, it is clear that certain concepts may be relearned—reweaved—in a dynamic process of rereading nature, women, humans in general, non-human life, as well as their relation to the all, or as claims Karen Ya-Chu Yang in her Introduction to this book, "how women are to reconnect with nature remains an ongoing challenge and prospect."

The majority of the chapters in this volume deal with ecofeminism with the understanding of such a complexity, but the connection between women and nature sometimes seems to resort to essentialist values. In this sense it is never too much to return to Diana Fuss's *Essentially Speaking* (1989), in that she suggests that the question of essence is grounded in history, philosophy, and politics, which point toward its character of immutability. This has been an immense and heated area of debate for feminism, whose end is not yet clear. All I want to say here is that Fuss's argument brings to the fore the contingence of the use of essentialism, precisely because of its historical character which opens up possibilities of changes—something that ecofeminism would address as a conceptual reweaving—because the sign is neither stationary nor uniform, let alone immutable. As a concept it needs renewal, for it is a productive term that the new generations of feminist and ecofeminist critics will still have to confront. I see this as a long journey to tread.[3]

In his discussion of Françoise D'Eaubonne, Luca Valera points to many relevant ideas, some of which bring the question of essentialism to the fore.

He refers to a non-dualistic anthropology which connects feelings and reason, feelings and will, a coexistence which, if peaceful, contributes to a harmonious relationship between humans, non-humans and nature. He also calls our attention to the ethics of care as the best effort offered by feminism and ecofeminism as it considers the complexity of the human being in our relationship to nature, and the need for women's "housekeeping nature." And yet, should this ethics be a prerrogative of women alone? Should it not be the task of all of us living in this planet?

As this Afterword was being written, I could not help thinking of the Samarco Dam break in Mariana, in the state of Minas Gerais, Brazil, which devastated the village of Bento Rodrigues. The irresponsible action of the company (Samarco) caused the environmental disaster for the Rio Doce river basin, as claimed Brazilian President Dilma Roussef during the Paris 2015 COP 21. It will take at least a decade for the complete recovery of the region. Obviously this is not a problem for women alone to solve, even if, as Valera says, they (we) are "the bearer of the emotional seed." I believe that all of us humans have to develop such a seed, but we also have to assure strong environmental laws that really defend the environment and preserve nature from such problems.

Adriana Teodorescu's chapter on motherhood presents new ideas on a difficult theme, be it in the West, North or South of the Globe. Despite the fact that there is an understanding of it being socially and culturally constructed, and that it is a biological non-choice, there are still many stereotypes associated with what Teodorescu describes as "the good mother paradigm." Yet she claims that the connection between women and nature is now seen in a different light, in that new representations of motherhood escape the ornamenting and idealized ones. For the critic, women involved in childbirth and mothering are "the agents that restore the ideological dominance of nature over culture." In this sense motherhood means empowerment for the woman, because she is to be seen as having "an ontological advantage over men."

Would this mean that women and men are still competing for superiority? Binary divisions are not helpful, despite the fact that the world still works under the oppressive dominance of patriarchal values. Should we not consider the need for a more supportive relationship that includes solidarity involving men and women? Such learning would help us understand the other (persons—human and non-human altogether), and live in a more egalitarian and harmonious way.

According to Teodorescu, "a natural component of femininity" is motherhood and breastfeeding. Yet although both are indeed something biologically associated with us women, identified with such gender, such perception seems to be problematic for those who do not take the concept of femininity essentially. As I read her views on motherhood, I thought of Elizabeth Grosz's (1990) problematizing of the concept of femininity. Her feminist discussion of the term via psychoanalysis may be helpful for our own understanding of the complexity of the concept. The characteristics of femininity, according to Freud's outline, include "seductive, coquettish behaviour, narcissism, vanity, jealousy, and a weaker sense of justice" (1990, 132), all of which are "strategies" for maintainance of the woman's status of object of desire, something that she manages to retain "through artifice, appearance, or dissimulation. Illusion, make-up, the veil . . . techniques she relies upon to both cover over and make visible her 'essential assets'" (132). This for Lacan, on the other hand, are attributes in the masquerade. Hence femininity for Grosz is seen as "practices," a perception that is closer to Freud, for whom the concept is seen as "techniques," rather than from Lacan. Such convergences and divergences—"practice" ("tecnique") is something one learns, whereas "attribute" seems to be inherent in the subject—still need to be better qualified in order for us to understand the concept and its complex use.

Breastfeeding and motherhood are a biological given, a non-choice, as claims Teodorescu, but I would not consider it as a qualifying term within the discussion of "femininity," for one cannot escape the idea of positionality, neither socially, nor culturally. Men and women can be in such positions, but still, motherhood will be always a non-choice for women. Femininity as a term needs a redefinition.

Anja Höing and Stephanie Baran discuss animals and women, and their association for different purposes with the help of ecofeminist's perspective of the interconnectedness of all oppressions. Höing's chapter analyses animal stories and her line of reasoning points to the use by different writers—male and female—of a recurrent pattern of oppression toward these animals, especially the female ones, that always have their roles minimized. Val Plumwood's problematizing of the theme helps in understanding such oppressive tints that show women and animals as the other.

Baran's essay analyses, in her turn, animals and women in an exchange of roles used by PETA to sell a commodity—the animals—by means of a "mantra": "sell women" to "sell animals." Such a paradoxical positionality reveals a controversial market strategics that works with ads in which women are used to frame oppression against animals. Some of the ads use celebrities and their bodies as commodities in order to sell the product. The connection discloses an ambivalent connotation: it can either promote empowerment for the woman, as Diana Villanueva Romero (2013) claims in her analysis of women and animals in fashion campaigns, or such a connection implies a body used as a "site of abuse" (166) when this likeness becomes derogatory, informing the oppressive charge channeled for both disempowered beings.[4]

The perception of animals and women as 'disempowered beings' leads to Anja's Koletnick's chapter on transgender individuals and the ecofeminist's understanding of meat non-consumption, diet, and body image. This chapter is an excellent reading for our understanding of ecofeminism and vegan feminism. Yet when Koletnick refers to diets as always having political connotations and implications, this made me think that the oppressive implications are much wider than the context of transgender individuals, for it includes race/ethnicity, gender, and class, alongside a whole complex of body patterns North, South, East, and West of the globe. The intersections can be endless.

Here, my illustration comes from literature and refers to the Caribbean poet Grace Nichols in her *The Fat, Black Woman's Poems* (1984). The body is for her a site of resistance and many of her poems express this perspective. It is contextualized as in opposition to oppressive dietary regimes by means of irony (Hutcheon 2005): the poem "Looking at Miss World," for example, discloses a fat black woman as the possible winner of a beauty contest. The skinny contestants are framed within a TV set, and, in the meantime, while at home watching the contest, the fat black woman celebrates her victory. Obviously what Nichols implies in her poem is the different Caribbean discursive context in which this woman appears; that is, outside the global Western perspective there are other body patterns that are not considered, and that such bodies are also "attached to the order of desire, meaning, and power" (Grosz 1994, 77). The body image presented in this case is not passive, even if one considers the questions of race/ethnicity as well as fat as present in the struggle shown in the aforementioned poem. Here what is also at stake is the question of otherness which testifies that the context is much wider, as already pointed out.

In Mitten and Amore's chapter, body images are the focus in terms of the "normative discontent," which leads to the perceptions of women's physical bodies as being internal and external. Unrealistic standards attached to women's bodies contribute immensely to the creation and introjection of negative images, which help in the development of low self-esteem, since not all women meet such standards. For the critics, "body dissatisfaction is present in the majority of women in developed countries, including women at normal weights." This is synonymous with depression, anxiety, social withdrawal, and isolation. Feminist and ecofeminist criticism deal with this issue in many ways. I thought of Debora Slicer's still valid claim: "We are a culture generally deaf to both our bodies and the rest of material life, deaf at an increasing cost" (1998, 61), for we do not listen to the body's speech. In other words, women seem to pay more attention to what culture forces them to accept at the cost of their (our) own health. Eating and other health disorders still cause women's somatizing symptoms in their (our) bodies. Culture regulates what is or is not possible within its oppressive normative standards, as argues Butler (2004).

And here, once more literature helps us understand culture: whilst Nichols's fat black woman is happy with her large body, the poem's profound sense of irony shows the lack of critical awareness in those who are taken as a kind of 'hostage' of culture (illustrated by the TV beauty contest), for they seem unable to see that the imposition of standards such as the beauty/fashion complex (responsible for the "tyranny of slenderness," to use Bordo's [1993] term) is oppressive for human beings—women, men, transgender individuals, and that it does not matter whether they are white, black, slim, fat, or vegan.

Other examples from literature come from two Brazilian contemporary writers whose novels unveil the question of body perceptions directed to a heterotopian context. Both discuss body images as related to the female protagonists and to those associated with them, including a transgender individual. Such is the case of *Pérolas absolutas* (Pure Pearls), by Heloisa Seixas

(2003). The heterotopic body belongs to a transgender individual, a "woman-fish" (that is her main feature), who occupies both male or female positions, depending on the client, in this case a heterosexual woman biologist. The other novel, *A chave de casa* (*The House in Smyrna*), by Tatiana Salem Levy (2007), problematizes the heterotopic question as related to the paralized body of a writer who only manages to heal herself as she goes to Turkey, and while there, she looks for her identity, and that includes a meeting with a Muslem girl in a Turkish Bath. Both novels disclose the body question as related to cultural problems of self-esteem, gender, and transgender identities. The place/space occupied by the body is a question related to affect and to transforming one's own body image into an empowered territory.

Such empowerment is immensely relevant be it in fiction or promoted by the real world. Mitten and Amore's chapter talk about the danger that internalizing negative body images may cause to women, for it leads to a disconnection with values that are greater than the body, referring, among other things, to spiritual values developed through women's positive relationship with nature. They talk about how being outdoors in nature has helped many women develop a sense of empowerment, because of what such interconnection means: "from microbes . . . to the sun on whose light we depend." This also means nature's allowing for differences—big and small. This, they say, helps in self acceptance, for the world is more than human.

Christina Holmes's chapter is also an incredible lesson in this respect, especially when she discusses the question of consciousness raising as related to Latina women immigrants in the Women Intercultural Center (WinC) that welcomes women immigrants from Mexico. The WInC welcomes these women in an affectionate, socially and politically responsible way, for it not only welcomes but helps them to become empowered in what Holmes calls "borderland ecofeminism," deriving the term from Gloria Anzaldúa's borderlands/*la frontera*. From her chapter I would like to mention her reference to Paulo Freire's "Pedagogia do Oprimido" (Pedagogy of the Oppressed), in which the educational practice is considered in relation to its political features that involve the oppressed in their own liberation from oppression, as well as their building up confidence in order to look after themselves, and the environment: "to raise ecological consciousness is to make it political." The WInC also deals with women's sense of belonging in its "affective and political aspects." Affect here is also related to women's health, a "multifaceted and socially constructed arena of concern" in which healing takes up many forms, which recalls again Slicer's (1998) perception about the unheard discourse of the body. Holmes refers to the interviews conducted with women at the WInC—immigrants and workers—and the responses dialogue with the other chapters in this book, especially because they go beyond the essentialist connection between women and nature already discussed in this Afterword. Holmes sees the responses as disconnecting women's bodies from nature, since they are seen "not as 'nature' but contested social territories, bodies that reveal and rework how power inscribes and subjugates," also that there's no "'natural' or biological link between women and nature." Hence it is useful here to understand that indeed the 'connection' between women and nature is far more complex than one might think. The body is a border, or as Anzaldúa quoted in Holmes's chapter points out, is "*una herida aberta*" (an open wound) to be healed in different ways, and ecofeminism can help in a variety of ways, one of them illustrated by the social and political work developed within WInC.

Holmes also discusses social action within the WInC, even though the Center is not directly involved in ecofeminist activism. Its action can clearly be associated to a political activism of a kind, in that far more than just welcoming women immigrants, its work toward consciousness raising opens up possibilities for these women to find their own independence in a country foreign to their experience in a whole range of respects, including language, and culture. The analysis is connected with the other chapters that discuss ecofeminist activism in relation to life narratives. Apart from the idea of activism most of the chapters call attention to the question of hope, a feature that I see as relevant for us eco/feminist readers today.

From Shlomit Tamari's chapter on the work of Israeli activist Bilha Givon more than anything else, I would like to stress the idea of perseverance behind the activist's practice, for not even a severe illness developed during her work prevented her from defending her ideas. Curiously enough, more than gender, her role as a fighter for environmental justice counted more than her being a woman activist, for her actions defined her as a "source of values," as Tamari points out. I believe that such a perception reinforces the question about what really matters: whether the awareness that (environmental) justice can be effected despite one's gender, or if one's gender has always to be the focus. By the same token, certain fights can only be fought by the oppressed themselves. This is what Jana Sawicki (1994) claims in her feminist approach to Foucault, often accused of not having a real interest in the feminist struggle. Bivon was also criticized for doing what she did—her activism was both outside and inside the industrialist system, becoming what Tamari claims as Sartre's dialectic of the third party. Being a soloist defines Bivon's life struggle as important but teaching us a lesson: individualism may lead to an unconcluded journey and environmental justice is a collective fight.

The question of the collective as opposed to individual fight for autonomy is present in Valerie Padilla Carroll's very perceptive and critical essay of the feminist magazine *Country Women*, in relation to the 1970s movement toward the back-of-the-land participants. Having a connection with the 19th-century nature writing which became popular in the beginning of the 20th century and reached maturity in the mid-1970s, the magazine called for a retreat into the countryside as a way of escaping from problems of different sorts brought by society: "a potent fantasy for many"; "idyllic narratives wrapped in a 'how-to' of self-sufficient agrarian living." This was critically viewed as "Fantasies of independence and autonomy" which had to do with "masculine endeavors" to be better supported by the women who were to accompany those who participated in the movement.

I thought of a Brazilian writer and educator, Júlia Lopes de Almeida, who lived during the first half the 20th century, and who wrote educational essays as well as novels addressed to women. She is one of the Brazilian first feminists, but her kind of feminism is now understood as a "possible feminism," considering the historical and cultural context in which she lived. I put it between inverted commas because in a sense it seems to follow the same line as the writing of *Country Women* as it "defends" women's supportive role for their partners. Almeida's thought defended women's education so that they could better organize the home and educate their siblings within the contraints of the patriarchal family, for it allowed women to be educated but they would not be freed for higher flights (Brandão 2002).

Padilla Carroll's analysis points to the maturity achieved by the back-to-the-land movement, but it continued the oppression of women masking it with solidarity and companionship. The idea of "empowerment" for the women is instead a mask for oppression, for they were only a supportive hand for their male partners. As a whole, the objective of the magazine meant women writing as a possibility of independence. Yet the magazine had the bias of the Western, white and middle-class (North) American woman as the pattern, and oppressive ideologies go unchallenged, and even women play the role of oppressors themselves.

The final point as refers to nature in relation to Padilla Carroll's chapter is that although it is a source of empowerment, again the Western is the structural beneficiary. Nature is also a "sacred goddess and feminist project," again strengthning the essentialist association between women and nature. The magazine cover enacts such association which, according to Padilla Carroll, can be viewed as an empowering for the woman and a liberatory message. Still she is not hopeful with such an image for there is "tension between the desires of individual autonomy and feminist collective sisterhood." The author's critical reading claims that the women seek but do not find connection. Hence the fight seems flawed.

Reena Dube's chapter presents a critical analysis of the use of documentary films, especially as the ones produced by women filmmakers in the Narmada Bachao Andolan (NBA) action. Her theoretical discussion claims that women filmmakers are more innovative and instructive when it comes to this kind of film, for they "elevate the film from a personal record into a political documentary." Her argument can in a way be contrasted with Tamari's chapter on Bilha Givon's kind of activism, for the main 'actresses' in the NBA documentary—the writer Arundhati Roy and the social activist Medha Patkar—have different performances. If Givon and Patkar are not really 'celebrities' of social activism, for their performance is a low profile one, Roy, on the other hand, plays the great Diva in the *Dam/Age* documentary, and wants to be in that place, hers is a "global ambition" as opposed to Patkars's "militant particularism" and almost invisibility in the documentary. Yet Dube defends Roy from having an opportunistic role in the film, for the writer seems to blur art and activism.

R. Sreejith Varma and Swarnalatha Rangarajan's chapter on life narratives of three women ecoactivists take me to my final comment. I would like to point to the fact that practice is indeed what may transform life into meaning, pending on the direction taken by such practice. The chapter poses the idea of life stories as the language of emotion and that the data collected is unrevealed by the statistics. Emotions such as discord, fear, anxiety are seen as "driving forces behind environmental ethic and political action." It is here that "human subjects and the environment become entangled." The life narratives by these women grassroot ecoactivists are stories about a kind of activism that is 'lived' rather than merely 'spoken,' as the authors claim. In their activism the ethics of Earth democracy leads one to find echoes of a hopeful future even if the context of protest in which the stories were enacted was a hopeless one: displacement, survival, and health for both humans and non-humans. The awareness of this kind of struggle is far greater than merely fighting big companies such as Coca-Cola, or the use of pesticide, for what the stories really defend is the need one has to feel (and be) in one's place. It is what Val Plumwood (2002) calls "place-sensitivity" (233), when it becomes possible to claim both critical and emotional approaches to space/place, or the women's "desire to restore their oikos," as claimed by the authors.

As a final point I would like to return to the initial idea of this Afterword being in a heterotopian place, a place that mediates between a conflicting yet possible dialogue that interweaves, interconnects, and intersects different thoughts and arguments, which expose the dense and complex world that nature, is alongside all its creatures, human and non-human alike. The chapters in this book are a challenge for all readers who want to deepen their understanding of the complex web of authors thinking ecofeminism. The connection between women and nature is well woven from the question wisely established since the book title, which takes us to the idea of intersectionality which weaves the chapters together, for, as Murphy (2011, 147) claims: "intersectional analysis becomes increasingly important across the widening range of ecofeminist concerns. And such analysis requires an ever more sophisticated inclusiveness in our rhetorical constructs and our ethical purviews. Such analysis . . . necessarily reject[s] dualistic and dichotomous thinking."

Hence, the chapters provide food for our thinking nature from different ecofeminist perspectives: from essentialism through activism; from oppressive to liberatory attitudes/actions toward what we do to our body image; to our life stories in relation to humans and to nature others; in relation to us as humans to the more than human in the world and the planet as a whole. Such thinking should help leading us to a perception of nature as empowerment, and a relation which means inclusiveness as well as an affectionate and politically responsible attitude toward the all. An ecofeminist's approach to nature as a process can teach us to dive into learning a new language. This book is part of this process.

Notes

1 See Brandão (2015) for a more in-depth feminist analysis of heterotopian places/spaces as related to two Brazilian contemporary women writers, Heloisa Seixas and Tatiana Salém Levy.
2 Although I am well aware of the current use of the term 'feminist ecocriticism' as interchangeable with ecofeminism (Gaard et al. 2013; Opperman 2013), I respect the choice made by the authors for ecofeminism (or eco-feminism).
3 This argument was presented by Brandão et al. (2015) in a plenary session of *XVI Seminário Nacional Mulher e Literatura/VII Seminário Internacional Mulher e Literatura*, held in Caxias do Sul/RS, Brazil, September, 14–16. The title of the work read is: "Traduções da cultura: para onde caminha a crítica feminista?" ("Translations of Culture: Where Does Feminist Criticism Go?" and had the collaboration of Ildney Cavalcanti, Ana Cecília A. Lima, and Claudia de Lima Costa, Brazilian feminist scholars from Federal Universities of Alagoas, and Santa Catarina, involved in a research project on feminist translations into Portuguese. For this Afterword I have translated the argument into English adapting the ideas as needed.
4 In the anthology *Literature and Ecofeminism* (Brandão forthcoming), my chapter discusses the poem "Hottentot Venus," by Jackie Kay, in which such a connection leads to the devaluation of the body of a black woman.

References

Alaimo, Stacy. 2000. *Undomesticated Ground: Recasting Nature as Feminist Space*. New York: Cornel University Press.

Armbruster, Karla. 1998. "'Buffalo Gals, Won't You Come Out Tonight?' A Call for Boundary-Crossing in Ecofeminist Literary Criticism." In *Ecofeminist Literary Criticism: Theory, Interpretation, Pedagogy*, ed. G. Gaard and P. D. Murphy. Urbana and Chicago: University of Illinois Press.

Bakhtin, Mikhail. 1984. *Rabelais and His World*, trans. Hélène Iswolski. Bloomington: Indiana University Press.

Bauer, Dale M. and S. Jaret Mckinstry, eds. 1991. *Feminism, Bakhtin and the Dialogic*. Albany: SUNY Press.

Bordo, Susan. 1993. *Unbearable Weight: Feminism, Western Culture, and the Body*. Berkeley: Berkeley University Press.

Brandão, Izabel, et al. 2015. "Traduções da cultura: para onde caminha a crítica feminista?" ("Translations of Culture: Where Does Feminist Criticism Go?"). *XVI Seminário Nacional Mulher e Literatura/VII Seminário Internacional Mulher e Literatura*, held in Caxias do Sul/RS, Brazil, September, 14–16.

Brandão, Izabel. 2002. "À Margem ds Alagoas: retratos de uma pesquisa." In *Retratos à Margem: antologia de escritoras das Alagoas e Bahia (1900–1950)*, ed. I. Brandão and I. Alves. Maceió: Edufal.

——. 2015. "Lugares heterotópicos e a constituição de corpos fronteiriços e identidades transitórias na narrativa de autoras contemporâneas." In *Espaço e gênero na literatura brasileira contemporânea*, ed. R. Dalcastagnè, V. M. V. Leal. Porto Alegre: Zouk.

——. (forthcoming). "Grace Nichols and Jackie Kay's Corporeal Black Venus: Feminist Ecocritical Realignments." In *Literature and Ecofeminism*, ed. D. A. Vakoch and S. Mickey.

Butler, Judith. 2004. *Undoing Gender*. New York: Routledge.

Foucault, M. 1984. "Of Other Spaces: Utopias and Heterotopias," trans. J. Miskowiec. In *Architecture/ Mouvement/Continuité* October. http://web.mit.edu/allanmc/www/foucault1.pdf

Fuss, Diana. 1989. *Essentially Speaking. Feminism, Nature, and Difference.* New York and London: Routledge.

Gaard, Greta and Patrick D. Murphy, ed. 1998. *Ecofeminist Literary Criticism: Theory, Interpretation, Pedagogy.* Urbana and Chicago: University of Illinois Press.

Gaard, Greta, Simon C. Stock, and Serpill Oppermann, eds. 2013. *International Perspectives in Feminist Ecocriticism.* New York and London: Routledge.

Grosz, Elizabeth. 1990. *Jacques Lacan—A Feminist Introduction.* London and New York.

——. 1994. *Volatile Bodies.* Bloomington: Indiana University Press.

Hutcheon, Linda. 2005. *Irony's Edge.* New York: Routledge.

Levy, Tatiana Salém. 2007. *A chave de casa.* Rio de Janeiro: Record.

Murphy, Patrick D. 2011. "Afterword." In *Ecofeminism and Rhetoric—Critical Perspectives on Sex, Technology, and Discourse*, ed. D. A. Vakoch. New York and Oxford: Berghahn Books.

Nichols, Grace. 1983. *I Is a Long Memories Woman.* London: Karnak House.

Opperman, Serpill. 2013. "Feminist Ecocriticism: The New Ecofeminist Criticism," *Feminismo/s* 22(diciembre): 65–88.

Plumwood, Val. 2002. *Environmental Culture: The Ecological Crisis of Reason.* London: Routledge.

Romero, Diana Villanueva. 2013. "'Savage Beauty': Representations of Women as Animals in Petas's Campaigns and Alexander McQueen's Fashion Shows," *Feminismo/s* 22(diciembre): 147–175.

Sawicki, Jana. 1994. "Foucault, Feminism and the Question of Identity." In *The Cambridge Companion to Foucault*, ed. G. Gutting. Cambridge: Cambridge University Press.

Seixas, Heloisa. 2003. *Pérolas absolutas.* Rio de Janeiro: Record.

Slicer, Débora. 1998. "Toward an Ecofeminist Standpoint Theory: Bodies as Grounds." In *Ecofeminist Literary Criticism—Theory, Interpretation, Pedagogy*, ed. G. Gaard and P. D. Murphy. Urbana and Chicago: University of Illinois Press.

Index